T0300983

DMAIC

The industry is currently facing a multitude of challenges on production lines that must be systematically addressed in order to improve quality and productivity. One commonly utilized methodology by companies is DMAIC (Define, Measure, Analyze, Improve, and Control), which has been proven effective through various case studies and is now considered a standard problem-solving approach. However, existing books on DMAIC often focus solely on defining concepts and tools, as well as showcasing case studies separately. This book consolidates all of these aspects into a single, user-friendly resource.

DMAIC: Concepts, Tools, and Industrial Applications offers both theoretical and practical insights into the DMAIC methodology. It sequentially presents all stages of DMAIC, outlining activities, data collection, and analysis processes. By integrating DMAIC concepts, tools, and case studies, this book provides industry professionals with a comprehensive resource. Real-world case studies from companies that have successfully implemented DMAIC to enhance production processes are showcased, offering valuable insights and practical examples.

This book serves as a valuable resource for professionals, students, and individuals responsible for improving industrial production systems. By consolidating theory, tools, and case studies into a single book, readers can gain a comprehensive understanding of DMAIC and its application in industry.

DMAIC

Concepts, Tools, and Industrial Applications

Arturo Realyvásquez Vargas, Jorge Garcia Alcaraz, Suchismita Satapathy, José Roberto Díaz-Reza, and Bogart Yail Marquez Lobato

CRC Press
Taylor & Francis Group
Boca Raton London New York

CRC Press is an imprint of the
Taylor & Francis Group, an **informa** business

First edition published 2025
by CRC Press
2385 NW Executive Center Drive, Suite 320, Boca Raton FL 33431

and by CRC Press
4 Park Square, Milton Park, Abingdon, Oxon, OX14 4RN

CRC Press is an imprint of Taylor & Francis Group, LLC

ISBN: 978-1-032-91280-6 (hbk)
ISBN: 978-1-032-93148-7 (pbk)
ISBN: 978-1-003-56460-7 (ebk)

DOI: 10.1201/9781003564607

Typeset in Times
by Apex CoVantage, LLC

Contents

Preface

The Define-Measure-Analyze-Improve-Control (DMAIC) methodology, a widely recognized and extensively used problem-solving approach in manufacturing systems, holds significant academic, industrial, and scientific importance. Its versatility is evident in over 5,000 publications (as of December 2023) exploring its application in diverse areas such as defect reduction, lead time management, inventory control, and production enhancement.

The DMAIC is composed of five phases. The first is the Define phase, which is crucial for the success of any process improvement project. Here, the problem to be solved is defined, the project's objectives are established, and the key performance parameters that will be used to measure success are identified. In addition, the stakeholders involved are identified, and a communication plan is established to keep all interested parties informed about the project's progress. In this phase, a solid working team is also defined, and the roles and responsibilities of the members are assigned because a well-organized and committed working group is essential for completing the subsequent phases of DMAIC.

The second phase is the Measure phase. During this stage, data are collected on the current process performance, which provides an objective basis for decision-making. Collecting and analyzing data in this phase allows the identification of areas of opportunity, the determination of the magnitude of problems, and the establishment of clear metrics to measure progress during subsequent phases of the project. In the third phase, Analyze, a comprehensive analysis of the data collected in the Measure phase was performed to identify the root causes of problems or defects. This analysis allows us to understand the relationship between variables and determine the factors contributing to the identified problems. In addition, the Analyze phase provides key information to make informed decisions about the corrective and preventive actions needed to address problems and improve process performance.

The next phase is Improve, in which tangible process improvements are made using data analysis tools and techniques to identify the root causes of problems and develop effective solutions. The proposed solutions are implemented and monitored during the Improve phase to evaluate their effectiveness. Finally, in the Control phase, measures and controls are implemented to ensure that the improvements made during the improvement process are sustainable and maintained over time. This involves establishing performance metrics and monitoring and developing action plans to address deviations or problems.

Given DMAIC's importance in the industry, this book provides a conceptual framework for the methodology, describes the auxiliary tools in each of its phases, and reports two case studies. It is divided into four chapters, as described in what follows.

Chapter 1 provides an overview of the DMAIC. This includes the methodology's concept and a detailed description of its five stages. It also mentions the benefits of DMAIC implementation to companies and the barriers and critical factors for its implementation. Finally, a bibliometric review shows the trends in DMAIC publications, the countries, institutions, authors, and journals that publish the most on this methodology, and the most cited authors, documents, and journals.

Chapter 2 provides a detailed description of 18 auxiliary tools that can be used in each DMAIC phase. These tools include SIPOC, VOC, stakeholder analysis, project charter, process flow diagram, data collection techniques, process mapping, KPIs, cause-and-effect diagram, statistical analysis, ANOVA, brainstorming, FMEA, DOE, pilot programs, control charts, SOPs, and Poka-yoke. The description includes each tool's concept and the advantages and disadvantages offered by their application.

Chapter 3 presents a case study on an assembly line producing 31–734 defective parts per week, with a dead time of 8.4 to 25.1 hours. The application of the DMAIC methodology resulted in a significant 67.3% reduction in defects, and the downtime was successfully reduced to less than five hours per week, showcasing the tangible benefits of this approach.

Chapter 4 includes the second case study, which deals with a company that, due to staff shortages during the COVID-19 pandemic, is unable to meet customer demand at normal times and is forced to work overtime. With the DMAIC methodology, the assembly line was redesigned, production was increased to 58 pieces/day, and demand was now 100% satisfied, eliminating overtime, which reduced production costs.

About the Authors

José Roberto Díaz-Reza has a master's degree in industrial engineering with a specialty in quality and a Ph.D. in innovation in product engineering and industrial processes from the University of La Rioja (Spain) and also a Ph.D. in advanced engineering sciences from the Autonomous University of Ciudad Juárez (Mexico). He has published a book and some articles in indexed journals related to lean manufacturing tools. Dr. Díaz-Reza has also participated in international congresses, and his area of interest is optimizing industrial processes using Lean Manufacturing tools.

Jorge Luis García Alcaraz is a full-time researcher in the Industrial Engineering and Manufacturing Department at the Autonomous University of Ciudad Juarez in Mexico. He received a Ph.D. in Industrial Engineering Sciences from Ciudad Juárez Technological Institute (Mexico), a Ph.D. in Innovation in Product Engineering and Industrial Process from the University of La Rioja (Spain), a Ph.D. in Sciences and Industrial Technologies from the Public University of Navarre (Spain), a Ph.D. in Mechanical Engineering from the University of Zaragoza (Spain). Dr. García is an author/coauthor of 299 documents indexed in Scopus and has attended more than 150 international conferences. His main research areas are the multicriteria decision-making process and techniques applied to lean manufacturing, production processes, and supply chain modeling.

Bogart Yail Márquez Lobato is a full-time Professor-Researcher at the Instituto Tecnológico de Tijuana and a member of the Sistema Nacional de Investigadores (National System of Researchers). He completed postdoctoral studies at the Colegio de la Frontera Norte and earned a Doctor of Science degree from the Universidad Autónoma de Baja California. Additionally, he holds a master's degree in applied economics and has a computer engineering background. His areas of interest include Social Simulation, Neuroscience, and Artificial Intelligence applied to social issues.

Arturo Realyvásquez Vargas is a full-time professor in the Department of Industrial Engineering at Tecnológico Nacional de Mexico/Instituto Tecnológico de Tijuana in Mexico. He received a Ph.D. in Engineering Sciences from the Autonomous University of Ciudad Juarez in Mexico and a Ph.D. in

Innovation in Product Engineering and Industrial Process at the University of La Rioja (Spain). Dr. Realyvásquez is an author/coauthor of 20+ papers published in journals indexed in the Journal Citation Reports and has supervised more than 20 bachelor theses and five master theses. His main research areas are industrial process optimization, lean manufacturing, and ergonomics.

Suchismita Satapathy is an associate professor at the School of Mechanical Sciences, KIIT University, Bhubaneswar, India. She has published over 130 articles in national and international journals and conferences and has more than 15 years of teaching and research experience. She has guided many Ph.D. (two completed, four ongoing), MTech (20), and BTech (49) students. Dr. Satapathy has published two books, and her research interests include production operation management, operation research, acoustics, sustainability, and supply chain management.

1 Define-Measure-Analyze-Improve-Control (DMAIC)—An Overview

1.1 CONCEPT OF DMAIC

The DMAIC methodology, short for its five phases: Define, Measure, Analyze, Improve, and Control, is a structured and effective tool to improve processes and achieve more consistent and efficient results continuously (Navarro-Romero et al. 2022). Motorola developed this methodology in the 1980s as part of the Six Sigma approach for process improvement (Cesarelli et al. 2021) and the implementation of long-term solutions to manufacturing (Aichouni et al. 2021). Subsequently, the DMAIC methodology became popular and an integral part of Six Sigma, widely used in many industries to improve process quality and efficiency significantly (Socconini 2023).

The DMAIC methodology helps identify, eliminate, and control product and production process quality problems that may arise (Khan et al. 2022). It aligns with lean manufacturing (LM) thinking principles and Kaizen, which promote continuous organizational improvement (Khan et al. 2022). DMAIC has proven to be an effective tool for continuously improving the production processes in various industries (Khan et al. 2022). For example, in the automotive sector, Toyota has been recognized for its successful implementation, which has allowed it to reduce its energy consumption by 50% and waste generation by

DOI: 10.1201/9781003564607-1

90%. However, it is essential to note that the use and application of the DMAIC methodology are not limited to the automotive industry, as they are currently widely used in various fields and sectors to achieve continuous improvements in processes and more consistent and efficient results (Daniyan et al. 2022).

In conclusion, the DMAIC methodology is paramount for process improvement, as it provides a structured and systematic approach to identifying, analyzing, and solving problems in manufacturing (Aichouni et al. 2021). In addition, the DMAIC methodology allows for measuring and controlling implemented changes to ensure long-term sustainable improvements in process efficiency and quality (Khan et al. 2022). In addition, the DMAIC methodology also helps to minimize variability in processes and to make decisions based on data and facts rather than assumptions or intuition (Velu et al. 2021). Properly implementing the DMAIC methodology can generate significant results in terms of efficiency, quality, and performance in an organization in any industry (Al-Qatawneh et al. 2019; Knop 2019). The DMAIC methodology also promotes a culture of continuous improvement in the organization, constantly reviewing processes, and searching for improvement opportunities (Velu et al. 2021).

Applying DMAIC requires a process that is divided into five stages, which are defined in what follows.

1.2 DMAIC PHASES

The DMAIC methodology comprises five distinct phases, each with its own purpose and specific activity (Dambhare et al., 2013). The first phase of DMAIC is "DEFINE." This phase is fundamental in the continuous improvement process as it allows one to clearly define and understand the problem to be solved, set the objectives to be achieved, and determine the metrics and success criteria to measure progress (Al-Qatawneh et al. 2019; Subagyo et al. 2020; Yuan et al. 2023). In this phase, all relevant information about the problem is collected, customer needs and expectations are identified, problem boundaries are established, and a multidisciplinary work team is created to carry out the project (Yuan et al. 2023). In addition, root cause analysis is performed to identify the main reasons for this problem (Godina et al. 2021).

Root cause analysis helps understand the internal and external variables that contribute to the problem, which allows the establishment of appropriate strategies and actions for its resolution (Godina et al. 2021). In addition, key performance indicators that will be used to measure project progress and customer satisfaction have been defined (Herrera Gómez and Trinidad Requena 2002). In summary, this phase focuses on clearly defining and understanding the problem, establishing the objectives and success criteria, gathering relevant information, identifying the client's needs, establishing the boundaries of the problem, and forming a multidisciplinary working team (Wibowo 2019).

The second phase is "MEASURE," where relevant data are collected to quantify the problem and establish a course of action (Al-Qatawneh et al. 2019). Potential causes of the problem are also identified, and how the success of the improvement is measured is defined (Cheng 2017). This phase focuses on data collection and measurement of the current process performance by analyzing the process characteristics and measures against established standards (Li et al. 2019; Subagyo et al. 2020; Syaifoelida and Ying 2020; Godina et al. 2021). Data can come from internal and external sources in the organization, and statistical tools are used to analyze and understand process variability (Syaifoelida and Ying 2020; Woodall et al. 2022). Therefore, the Measure phase of the DMAIC method is crucial for accurately understanding the current state of the process under analysis (Woodall et al. 2022). In addition, key performance indicators are established in this phase, and benchmarking is performed against the established standards (Niñerola et al. 2020). In summary, the Measure phase of the DMAIC method focuses on data collection and measurement of the current process performance (Alvarez-Rozo et al. 2020).

The third phase is "ANALYZE," the collected data are comprehensively examined, and statistical analysis tools are applied to identify patterns, trends, and relationships between variables (Dambhare et al. 2013). In addition, it seeks to identify the root causes of the problem and understand how they affect the process, and thus be able to implement improvement actions that eliminate or reduce these problems or defects in the process (Hors et al. 2012; Al-Qatawneh et al. 2019). These causes can be related to materials, methods, labor, equipment, or environment (Da Rocha et al. 2010). Statistical tools and techniques are used to collect and analyze the data for practical analysis. These tools and techniques include analysis of variance, process mapping, failure mode and effect analysis, and the Pareto diagram (Zdęba-Mozoła et al. 2023). During the analysis, the project team focuses on thoroughly understanding the variables influencing the process and determining which contribute to the problems or defects (Girmanová et al. 2017; Zdęba-Mozoła et al. 2023), which will make it possible to generate an order of priority that will make it possible to address them.

The fourth phase is "IMPROVE," where solutions are generated and implemented to address the root causes identified during the previous analysis phase (Subagyo et al. 2020). These solutions are evaluated, and the most effective one is selected to be implemented based on the company's own needs. During this phase, the data and information collected in the previous phases are analyzed to identify improvement opportunities and develop strategies to implement positive changes in the production process (Guo et al. 2019). The improvement team uses statistical tools and techniques to evaluate the root causes of the identified problems and conducts experiments or pilot tests to test possible solutions and their impact on the response variables (Yuan et al. 2023).

Once the most promising solutions have been identified, process changes will be implemented, success criteria will be established, and the impact of the implemented improvements will be measured (Al-Qatawneh et al. 2019). In addition, control plans are designed to ensure that the changes made are maintained in the long term and monitored to ensure continuous improvement (Yuan et al. 2023). During the "IMPROVE" phase, the improvement team must also integrate all involved individuals, departments, and stakeholders, such as employees, customers, and suppliers, to ensure that the proposed solutions are feasible and meet the needs and expectations of all stakeholders (Guo et al. 2019). In summary, the "IMPROVE" phase of DMAIC is crucial for identifying practical solutions and making positive changes to the process, focusing on improvement and, above all, with the customer (Yuan et al. 2023).

Finally, the fifth phase is "CONTROL," which focuses on monitoring and maintaining the improvements implemented to ensure the changes are sustained over time (Cheng 2017). During this phase, the necessary controls, monitoring, and feedback systems are implemented, and follow-up plans are developed to ensure that the improvements are sustainable and the stated objectives are met (Dambhare et al. 2013). In addition, those responsible for following up on the established parameters, generated reports, and reporting frequencies should be established.

Similarly, metrics are established and constantly monitored to evaluate the performance of the improved process and ensure that it meets the standards set in the previous stages (Li et al. 2019). In addition, action plans are developed in this phase to address any deviations or problems arising in the implementation process, including training initiatives or adjustments to standard operating procedures (Vasconcellos de Araujo 2020). The "CONTROL" phase also involves establishing clear responsibilities and allocating necessary resources to ensure the proper implementation and sustainability of the changes (Vasconcellos de Araujo 2020). In summary, the control phase in the DMAIC method is fundamental to ensure that the implemented improvements are sustained over the long term (Banawi et al. 2020).

1.3 BENEFITS OF DMAIC

Companies apply the DMAIC because they obtain a series of tangible or intangible benefits. Although it is associated with quality and problem-solving, many others exist. The most important are the following:

(1) Improving customer satisfaction: DMAIC focuses on customer satisfaction and, therefore, ranks first. It helps identify and rectify process inefficiencies, leading to improved product or service

quality, which undoubtedly represents higher customer satisfaction (Subagyo et al. 2020).

(2) Decrease in defects and errors: Because of its systematic approach, DMAIC helps reduce defects and errors, which improves the overall quality of products and services (Amrina and Firmansyah 2019).

(3) Economic advantages: Traditionally, DMAIC has been related to cost savings obtained by reducing waste, higher reproducibility according to quality standards, and a lower amount of red labor, indicating higher profitability for firms (Noori and Latifi 2018).

(4) Promoting continuous improvement: Because of its approach, DMAIC offers a structured approach to identifying inefficiencies, optimizing processes, and sustaining performance improvements over time (Cano et al. 2021). This represents a focus on continuous improvement at all structural levels of the company.

(5) Improving employee morale and communication: DMAIC is an approach that requires the integration of all people to solve problems. Therefore, involving employees in the improvement process and equipping them with structured tools and techniques improves morale, communication, and decision-making abilities, as they feel part of the business (Prashar 2014).

(6) Process quality improvement: Although DMAIC is traditionally associated with product quality, it is only the final variable, as it also involves improving production processes and is now a critical strategy for companies seeking operational excellence in their production lines (Hardy et al. 2021). (Hardy et al. 2021).

(7) Cost reduction: Implementing DMAIC leads to production cost savings, which improves a company's financial profit margin in the long run (Nagi and Altarazi 2017). However, it is worth mentioning that this initially requires considerable investment in employee training and dissemination.

(8) Facilitation of root cause analysis: Because of its structured approach, DMAIC facilitates root cause analysis in production processes, which makes it possible to identify and address problems that provide solutions to those problems that affect productivity, efficiency, and quality indexes of the processes (Velu et al. 2021).

(9) Increased efficiency: As mentioned previously, rationalizing resources through DMAIC in the processes eliminates waste, thus helping to improve operational efficiency and productivity (Kumar et al. 2021).

(10) Improving performance indicators: In the DMAIC stage, performance indicators should be identified, as this methodology is focused on improving them. Such indicators can be Z sigma, Cp, cycle time, and lead time, which contribute to overall process performance (Ahmed et al. 2023).

Apart from Toyota and Motorola, several companies and industrial sectors have applied DMAIC in their production and administrative processes, among which the following can be mentioned:

- A food producer in Noriega has improved environmental sustainability in the food processing industry, showing how process improvement can contribute to greater sustainability (Powell et al. 2017).
- Process defects have been reduced in the telecommunications industry, leading to improved quality and operational efficiency (Abhilash and Thakkar 2019).
- University sustainability has been improved using the DMAIC approach, demonstrating the versatility of the methodology across different sectors (Hamdan et al. 2024).
- A focused case study on additive manufacturing demonstrated that DMAIC improves the quality and sustainability of the production process (Rodriguez et al. 2022).
- In the glass industry, DMAIC minimizes defects and increases customer satisfaction with fewer rejected batches (Yadav et al. 2019).
- In an aluminum profile extrusion process established in Palestine, Lean Six Sigma was implemented using the DMAIC methodology to improve the process and quality and reduce waste (Araman and Saleh 2023).
- A case study of an automotive parts supplier showed how it reduced defects using DMAIC and Lean Six Sigma, which reduced customer rejections and challenges (Condé et al. 2023).

1.4 DMAIC BARRIERS

Implementing DMAIC methodology in production lines is difficult. Companies face several obstacles that can hinder successful implementation, and there are reports that these processes are frequently abandoned. Some of the most common challenges identified in the literature:

(1) Employees may resist change and adopt new processes and methodologies, especially if they are used in traditional working methods and are afraid to leave their status quo (Powell et al. 2017).
(2) The lack of top management support and commitment makes it difficult to drive organizational change and is associated with

low resource allocation, which can be challenging (Zakari 2014). Therefore, it is often stated that DMAIC should be initiated at the top management level to ensure commitment.

(3) Inadequate resource allocation may be due to a lack of commitment from top management or that no structure or work plan has been designed to implement DMAIC (Scheller et al. 2021). Therefore, the working groups and senior management responsible for resource allocation must prioritize the projects' importance so that an order can be established to find solutions to the problems.

(4) Insufficient employee training prevents employees from knowing the statistical and management tools for the practical application of the methodology. (Galvão et al. 2022). The previous information indicates that a training program should accompany DMAIC during the employees' working hours and preferably never during overtime since it could be seen as an additional workload that they are often unwilling to accept.

(5) Difficulties in censusing and capturing information for adequate analysis to facilitate decision-making (Khan et al. 2019). This is often because information comes from different departments, making it difficult to flow smoothly. In addition, specialized technology is often required to monitor processes requiring senior management investment.

(6) Internal resistance, where trade unions have a very notorious presence among workers, particularly in developing countries, where any change or project must be analyzed and approved by them (Prashar 2014). For this reason, senior management must properly manage projects with these organizations within the company, letting them know that the changes are for the good of all and continuous improvement. It must ensure that when making changes, the physical integrity of people or their continuity within the company is not endangered.

(7) Lack of creativity focused on continuous improvement and innovative thinking can hinder the problem-solving and improvement efforts inherent to the DMAIC (Condé et al. 2023). This is mainly due to the lack of technical knowledge of production processes. However, this may also be because they are young companies that do not have enough experience in top management and employees; therefore, it is recommended that there is always an external coach to provide support in this situation.

(8) Operational inefficiencies prevent managers from focusing on continuous improvement and seek only to solve day-to-day problems. (Subagyo et al. 2020). This occurs when production processes are

not adequately standardized, and managers seek to solve day-to-day problems without focusing on long-term improvement.

(9) The lack of a Lean Six Sigma culture prevents the valuing of continuous improvement and data-driven decision-making, focusing on solving problems based on experience and empirically. (Trubetskaya et al. 2023). This may be due to a lack of knowledge of the necessary statistical tools to facilitate analysis and decision-making, which is solved through training courses at all levels.

(10) The complexity of production processes and organizational structures poses challenges to effectively applying DMAIC principles for improvement, as there are no well-defined roles within the company (Panayiotou and Stergiou 2023).

1.5 CRITICAL SUCCESS FACTORS OF DMAIC

Several barriers prevent the implementation of DMAIC from having adequate success in production lines, so it is important to analyze what critical factors must be present to ensure that the efforts made are not in vain. A critical success factor (CSF) in the industry is an essential element that, if met, significantly increases the chances of a company or project succeeding; that is, it is the key area where things must go right for the company to thrive in a competitive market.

Several studies have been conducted on DMAIC CSFs, and some authors have focused on particular ones. The most common CSFs are as follows:

(1) Top management supports and commits several reasons. They are responsible for showing leadership and a vision of what is desired with DMAIC (Albliwi et al. 2014). They are responsible for allocating resources and budgets for all projects to be implemented and are the first to support the removal of barriers within the company. This act of leadership makes employees take the culture of continuous improvement seriously, and they are also responsible for recognizing and celebrating the successes of work teams.

(2) Top management leadership and management (Näslund 2013). They know their vision and propose a company's strategic direction. They are also responsible for making decisions when integrating workgroups, selecting projects, and allocating resources.

(3) Organizational culture is another factor since when this is positive, a climate of trust and collaboration among employees is fostered

(Näslund 2013). Employees should feel comfortable sharing ideas, information, and opinions to improve the production process. Participative people are always more open to change, strengthening the culture of continuous improvement in the joint learning and development environment.

(4) The skills and experience of all senior management members ensure that they can provide leadership and direction for often complex projects to be solved (Näslund 2013). In addition, skills and experience are required to make effective decisions that can solve their problems and challenges. This will undoubtedly motivate employees to feel supported by someone who can help them. Finally, it is important to mention that within DMAIC, there will always be constant training and development for senior management and not only for employees.

(5) Management commitment and involvement to ensure leadership and strategic vision in each project (Kumar Sharma and Gopal Sharma 2014). In addition, top management is responsible for fostering communication between all departments and sensitizing people to the need for change across the organization, which removes many interdepartmental barriers. Similarly, highly committed managers allocate resources and support projects appropriately. At the end of the project, they will offer recognition and awards to those who have committed to their projects and obtain favorable results.

(6) Linking project objectives to customer requirements and business strategies (El Safty 2011); otherwise, DMAIC would lose its reason for being. This linkage ensures that the project focuses on what matters to the company and client, which helps avoid deviations and unintended results. In addition, this linkage allows prioritization of the project's objectives. It facilitates the measurement and evaluation of its success in determining whether the project meets its objectives. Thus, by creating this linkage, continuous improvement focused on the client is achieved, and new areas or opportunities to work on are identified.

(7) Effective interpersonal communication between all individuals and departments ensures the understanding and collaboration of the work teams (Alhuraish et al. 2017). Recall that DMAIC is based on collaborative work and high information sharing is required at all levels. In addition, problems that arise in one department are often repeated in another, which makes it easier to use the knowledge already acquired in other areas of the organization. It is also frequently stated that change management is facilitated throughout a company when there is interdepartmental collaboration.

(8) Employee training is required to use and manage standard tools in DMAIC, such as data analysis and problem-solving tools (Stankalla et al. 2018). In addition, such training and employee training motivate employees, as top management shows that they are willing to invest in their personal development. Moreover, such training always provides environments that encourage continuous improvement by promoting the search for alternative ways to develop jobs.

(9) Employees' working hours should be considered (Swarnakar et al. 2020; Samanta et al. 2024). Workshops and seminars should be planned during working hours. Otherwise, employees consider DMAIC an additional burden that does not contribute to their income.

(10) A step-by-step approach using the DMAIC methodology (Alhuraish et al. 2017; Maciel-Monteon et al. 2020). It is often indicated that top management and experienced employees may be motivated; however, the steps and development are not adequately followed. This often occurs when highly trained employees and leaders take for granted certain activities that the rest of the workgroup does not understand.

(11) DMAIC should be integrated into a continuous improvement framework (Prashar 2014). DMAIC should be part of a generic framework aligned with the strategic planning of which top management is aware. In other words, DMAIC is another tool that should be employed within the company's overall objectives.

(12) Integration with environmental considerations for sustainability, as there is now widespread concern about the environmental aspects of this type (Antony and Banuelas 2002; Prashar 2014). That is, projects should focus not only on improving the process from an operational, technical, or economic point of view but also on integrating aspects related to the environment and the social welfare of employees because solutions may be operationally adequate but hurt the environment or put the integrity of workers at risk.

1.6 DMAIC—A BIBLIOMETRIC REVIEW

Given the importance of the DMAIC methodology in the industry, a bibliometric review is crucial because it allows for a systematic analysis of the existing literature on this topic, providing information on trends, patterns, and gaps in the research related to this methodology (Karakose et al. 2021). Thus, those interested in DMAIC applications and concepts can identify the most influential publications, authors, and journals, which can help guide future research directions and collaboration (Jain et al. 2022). This can also help identify where the research groups and consolidated institutions in that line of research are located, making it easier for potential graduate students to choose where to continue their studies or stay.

Similarly, a bibliometric review helps assess the impact and effectiveness of DMAIC implementation in various industrial sectors. Bibliometric reviews on DMAIC have already been conducted, albeit with a particular focus; for example, studies have been reported to evaluate the application of Lean Six Sigma, which includes DMAIC, in different sectors such as healthcare and manufacturing, highlighting the benefits and challenges encountered during implementation (Souza et al. 2021; Kumar et al. 2021). When delving into the content, many of these bibliometric analyses refer to Six Sigma using the DMAIC methodology; therefore, they are sometimes somewhat generic.

The objective of this section is to report a bibliometric analysis that answers the following questions:

- What are the trends in publications regarding DMAIC?
- Which authors, institutions, and countries produce the most about DMAIC?
- Which journals published the most articles on this topic?
- Which authors, institutions, or countries are the most cited?
- What keywords are the most commonly used by authors and editors when publishing documents?

To achieve this, a document search was performed in the Scopus database using the following search equation: TITLE-ABS-KEY (((DMAIC OR "Define—Measure—Analyze—Improve—Control") AND TITLE-ABS-KEY (manufacturing))). The word DMAIC, in case the acronym was used in the title, abstract, or keywords, would be included in the analysis, while the set of words "Define—Measure—Analyze—Improve—Control" referred to the whole word and finally, the word "manufacturing" indicates that the documents should refer to practical applications in a productive process associated with manufacturing or transformation of raw materials and not to services. The following results were obtained based on the questions asked previously.

1.6.1 Trends in DMAIC Publications

A total of 699 documents were identified using the previously mentioned search equation, and Figure 1.1 illustrates the trend in the publication of documents for the period between 2001 and 2023. It can be seen that the first document

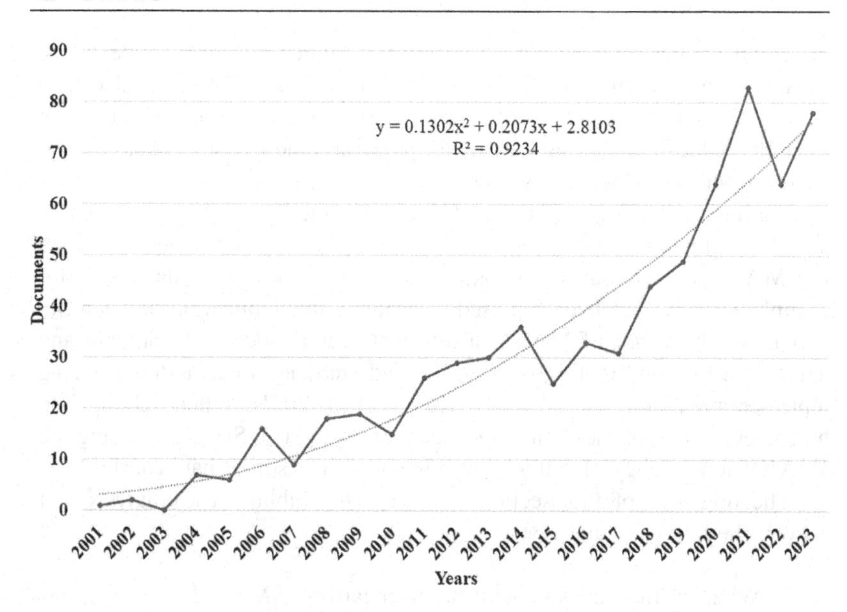

FIGURE 1.1 Trend of DMAIC publications.

about this methodology appeared in 2001, and Neuscheler-Fritsch and Norris (2001) were the first to do so. In 2002, two documents appeared, whose authors were Pavletić and Soković (2002) and Scaff (2002), all discussing Six Sigma as a production philosophy encompassing the DMAIC methodology.

In 2003, no documents were published on this topic, but from then on, the number increased steadily until 2014, when up to 36 documents were published annually. However, in 2015, there was a slight drop, with only 25 documents published. After that year, the changing intensity accelerated because constant growth was maintained until 2021 when 83 documents were published annually.

The dotted line indicates the trend adjusted by a quadratic function with a fit of 92.34%. This indicates that DMAIC will continue to be of academic and scientific interest since growth is generally observed.

1.6.2 Types of Documents Published on DMAIC

These documents can be of different types, and Figure 1.2 illustrates the distribution of these documents. It is clear that most of them are articles published in journals and, therefore, have undergone a peer review process, which guarantees their quality. The second category with the highest number of documents is conference articles, which also undergo a peer-review process but are less arduous or in-depth

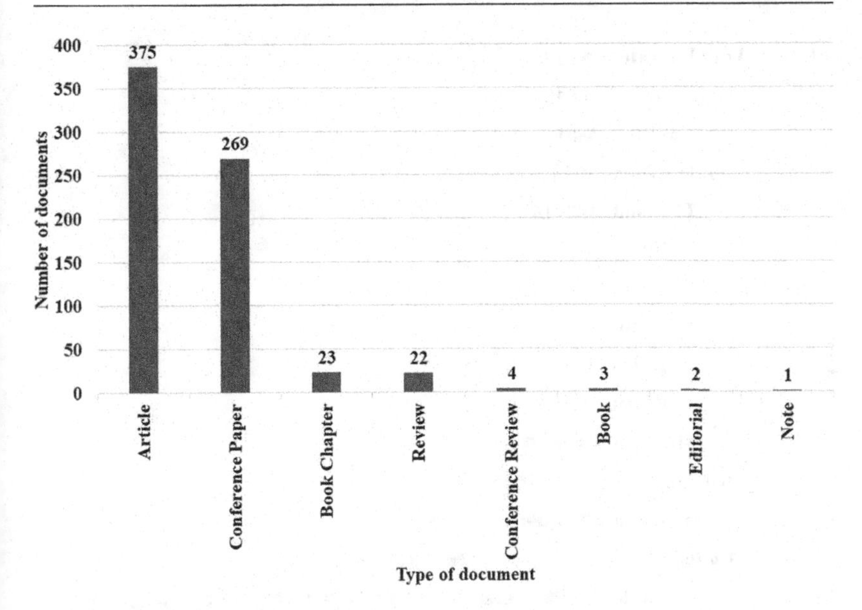

FIGURE 1.2 Type of documents published on DMAIC.

than articles in peer-reviewed journals. These two categories represented 92.13% of the total number of documents published during the period under analysis.

Other categories with fewer numbers refer to chapters in books and reviews, which can be metrics or literature bibliographies. Some books, book forewords, and notes by editors were also observed. Figure 1.2 indicates that most of the documents on this topic have been reviewed and are reliable.

1.6.3 Top Publishing Countries

Although the DMAIC methodology originated in American and Japanese companies, these are not always the countries that publish the most on this topic. In this case, 71 countries have been identified in which at least one document related to DMAIC has been published. However, it is important to mention that 15 of these documents were not associated with any specific country or territory.

Figure 1.3 illustrates the distribution of the countries that have published the most papers on at least ten DMAIC documents. India has the most significant number of documents, surpassing the United States in second place by more than 100%. Third, Indonesia is followed by the United Kingdom and Malaysia, which account for 60.80% of the total number of documents produced worldwide. It is important to note that three of the countries in second place are

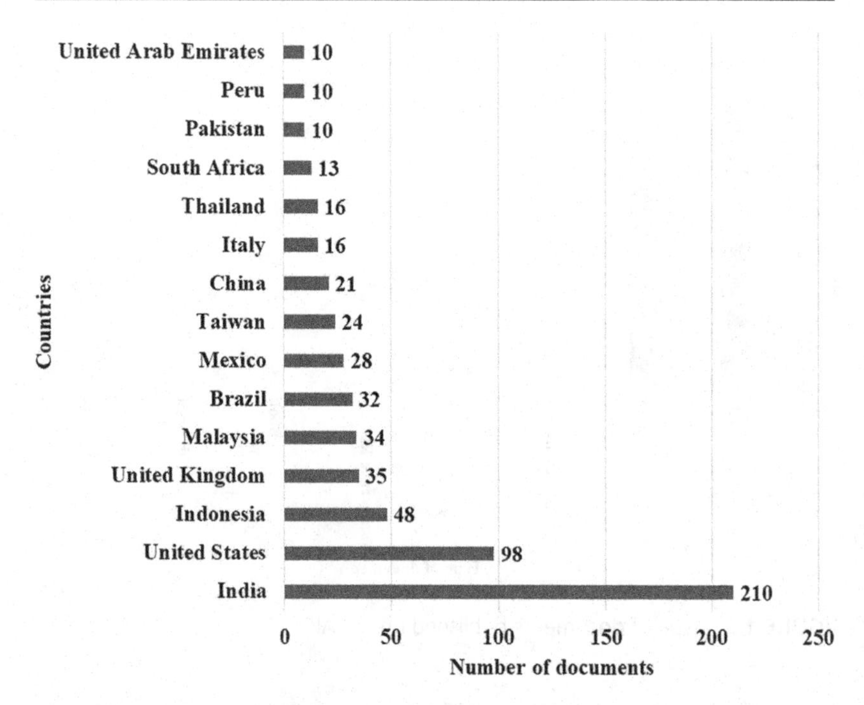

FIGURE 1.3 Countries that publish the most on DMAIC.

of Asian origin (India, Indonesia, and Malaysia), but it is also important to note that in the list of the 15 most productive countries, there are three Latin American countries, specifically Brazil, Mexico, and Peru, which indicates the level of academic and scientific importance of the DMAIC in these countries.

It was expected that European countries with very advanced manufacturing industries would produce more documents on this topic; however, only the United Kingdom and Italy appear on the list.

1.6.4 Institutions That Publish the Most

A total of 546 institutions were identified as having at least two papers published on the topic of DMAIC, indicating that another 153 institutions have published at least one paper. Figure 1.4 illustrates the top ten institutions that publish on DMAIC, where it is clear that the three institutions with the highest productivity are located in India: PSG College of Technology, Maharshi Dayanand University, and Indian Statistical Institute Bangalore. On this occasion, Tecnológico de Monterrey, established in Mexico, is in fourth place, with

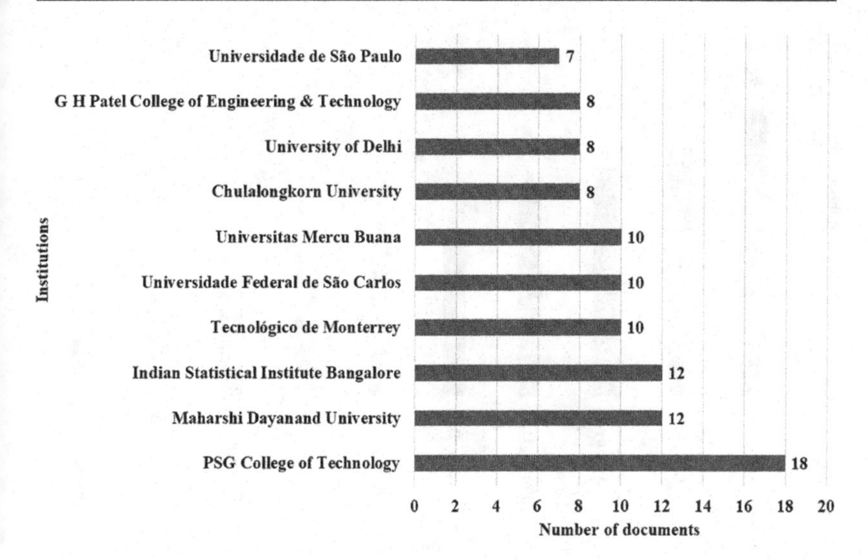

FIGURE 1.4 Institutions publishing on DMAIC.

ten documents produced. However, it is essential to mention that institutions such as the Universidade Federal de São Carlos in Brazil and Universitas Mercu Buana in Indonesia also have ten papers.

These results are in agreement with those reported in Figure 1.3, where India, Indonesia, Brazil, and Mexico are among the leading countries in which this topic of DMAIC has been studied. In other words, these six institutions had 10.30% of the documents generated on this topic, and it can be assumed that there are well-established research groups in these areas. Thus, students who wish to pursue a postgraduate degree related to Six Sigma and its methodologies can consult these institutions' educational programs because they will indeed have lines of research from which these publications arise.

1.6.5 Authors Publishing on DMAIC

A total of 679 authors were identified as having at least one published document on DMAIC; however, a few stand out, and Figure 1.5 illustrates those ten with the highest number. It is essential to mention that 11 authors were identified as having four papers, 36 as having three, 100 as having two, and the rest as having only one. When looking at the authors' surnames and reviewing their affiliations, it can be seen that many are located in countries such as India and Malaysia, which stand out in terms of the number of papers published.

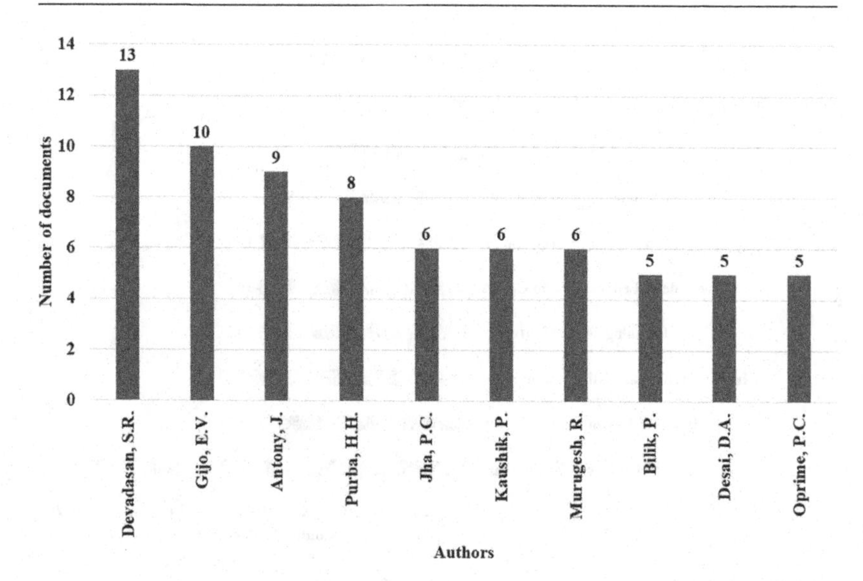

FIGURE 1.5 Most published authors on DMAIC.

1.6.6 Journals That Publish on DMAIC

A total of 324 journals were identified with at least one published document related to DMAIC; however, it has been detected that highly specialized journals deal specifically with this topic and, logically, stand out above the others. For example, the International Journal of Six Sigma and Competitive Advantage and the International Journal of Lean Six Sigma have led this topic.

Figure 1.6 illustrates the top ten list of journals, although it is worth noting that many of them refer to specialized conferences as well, such as Lecture Notes In Mechanical Engineering, IOP Conference Series Materials Science And Engineering, AIP Conference Proceedings, and Materials Today Proceedings.

It is essential to mention that four sources were identified with at least six papers: five with five, twelve with four, nine with three, 36 with two, and the rest with only one paper. Thus, scholars wish to publish their papers on Six Sigma, and the DMAIC methodology has several options, some of which are very specialized.

1.6.7 Most Cited Documents, Authors and Journals

Not all papers published on the DMAIC are equally cited; in fact, it is possible that very good papers published recently have fewer citations. Table 1.1 illustrates the top ten papers and their number of citations. Kumar et al. (2006),

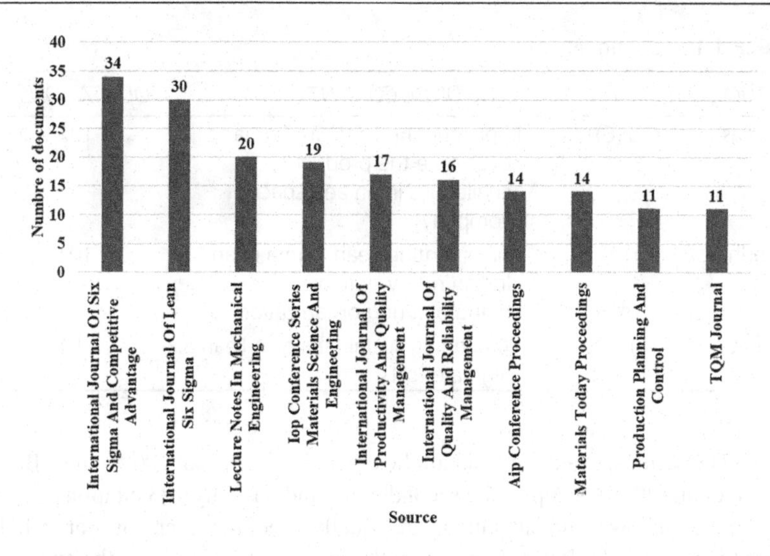

FIGURE 1.6 Sources that publish the most on DMAIC.

TABLE 1.1 Most Cited Documents on DMAIC

AUTHOR	*NAME OF DOCUMENT*	*QUOTATIONS*
Kumar et al. (2006)	Implementing the Lean Sigma framework in an Indian SME: A case study	371
Ben Ruben et al. (2017)	Implementation of Lean Six Sigma framework with environmental considerations in an Indian automotive component manufacturing firm: a case study	123
Chen and Lyu (2009)	A Lean Six-Sigma approach to touch panel quality improvement	121
Jirasukprasert et al. (2015)	A Six Sigma and DMAIC application for the reduction of defects in a rubber gloves manufacturing process	115
(Wang and Chen 2010)	Applying Lean Six Sigma and TRIZ methodology in banking services	110
Gijo et al. (2011)	Application of Six Sigma methodology to reduce defects of a grinding process	109
Swarnakar et al. (2020)	Deploying Lean Six Sigma framework in an automotive component manufacturing organization	104

(Continued)

TABLE 1.1 (Continued)

AUTHOR	*NAME OF DOCUMENT*	*QUOTATIONS*
Thomas et al. (2016)	Implementing Lean Six Sigma to overcome the production challenges in an aerospace company	102
Vinodh et al. (2014)	Implementing Lean Sigma in an Indian rotary switches manufacturing organization	101
Erdil et al. (2018)	Embedding sustainability in Lean Six Sigma efforts	91

with 371 citations, stand out from all the others by more than 200%, since Ben Ruben et al. (2017) occupies the second place and has only 123 citations.

In the same way, not all authors are equally cited; however, some are cited more than others. Table 1.2 shows the authors or groups of authors, the number of documents they have, the citations they have achieved, and the average number of citations per author. The data are arranged in descending order according to the number of citations obtained, where it is clear that authors from India are the most cited. It is essential to mention that when observing Tables 1.1 and 1.2, there are many coincidences among the authors, at least in the first three.

Table 1.3 shows the names of the most cited journals, the number of documents they generated, the citations obtained, and the average number of citations obtained for each document they published. This table shows that Production Planning and Control is not one of the most published topics on this topic; however, it is the one that has had the most citations in its 11 papers, while the International Journal of Lean Six Sigma, which occupied the first place, now occupies the second place with 30 published papers, and the same happens with the International Journal of Six Sigma and Competitive Advantage.

1.6.8 Most Cited Institutions and Countries

Table 1.4 illustrates the institutions or organizations that have received the most citations on the topic of DMAIC. Of the ten institutions, four belong to India, the country where most are published. There are also institutions from the United Kingdom and Taiwan, but a closer look at the documents shows that there are often joint productions between India and the United Kingdom.

TABLE 1.2 Most Cited Authors on DMAIC

AUTHOR	*DOCUMENTS*	*CITATIONS*	*AVERAGE*
Kumar M.; Antony J.; Singh R.K.; Tiwari M.K.; Perry D.	1	371	371
Ben Ruben R.; Vinodh S.; Asokan P.	1	123	123
Chen M.; Lyu J.	1	121	121
Srinivasan K.; Muthu S.; Devadasan S.R.; Sugumaran C.	3	121	40.33
Gijo E.V.; Scaria J.	2	120	60
Jirasukprasert P.; Garza-Reyes J.A.; Kumar V.; Lim M.K.	1	115	115
Wang F.-K.; Chen K.-S.	1	110	110
Gijo E.V.; Scaria J.; Antony J.	1	109	109
Swarnakar V.; Vinodh S.	1	104	104
Thomas A.J.; Francis M.; Fisher R.; Byard P.	1	102	102
Vinodh S.; Kumar S.V.; Vimal K.E.K.	1	101	101

TABLE 1.3 Most Cited Sources on DMAIC

SOURCE	*DOCUMENTS*	*CITATIONS*	*AVERAGE*
Production Planning and Control	11	1111	101
International Journal of Lean Six Sigma	30	714	23.80
Total Quality Management and Business Excellence	7	370	52.85
International Journal of Six Sigma and Competitive Advantage	34	250	7.35
International Journal of Quality and Reliability Management	16	242	15.12
International Journal of Advanced Manufacturing Technology	5	227	45.40
TQM Journal	10	222	22.20
International Journal of Productivity and Quality Management	17	203	11.94
Materials Today: Proceedings	14	178	12.71
Quality and Reliability Engineering International	3	164	54.66
Journal of Cleaner Production	2	152	76

TABLE 1.4 Most Cited DMAIC Organizations

ORGANIZATION	*DOCUMENTS*	*CITATIONS*	*AVERAGE*
Birla Institute of Technology, India	1	371	371
Glasgow Caledonian University, United Kingdom	1	371	371
Indian Statistical Institute, India	2	148	74
Nan Jeon Institute of Technology, Taiwan	1	121	121
Dublin City University, Ireland	1	115	115
University of Warwick, United Kingdom	1	115	115
National Taiwan University of Science and Technology, Taiwan	1	110	110
Nirmala College, India	1	109	109
Indian Statistical Institute, India	1	109	109
University of New Haven, United States	1	91	91

TABLE 1.5 Most Cited Countries on DMAIC

COUNTRY	*DOCUMENTS*	*CITATIONS*	*AVERAGE*
India	210	3204	15.26
United Kingdom	35	1357	38.77
United States	97	971	10.01
Taiwan	24	535	22.29
Italy	16	356	22.25
Malaysia	34	341	10.03
Brazil	32	291	9.09
Ireland	9	250	27.78
Portugal	7	145	20.71
Indonesia	48	135	2.81

Table 1.5 illustrates the top ten most cited countries, ordered in descending order by number of citations, with India, the United Kingdom, and the United States in the first place. However, if they were ordered according to the average number of citations per document, the United Kingdom would occupy the first place, followed by Ireland, whereas Indonesia would occupy the last category.

1.6.9 Most Used Keywords

A total of 1546 words were identified, and an analysis was performed on words that appeared or were repeated at least seven times. It was found that the word DMAIC appears 325 times: Six Sigma 313, Lean Six Sigma 94, Lean Manufacturing 54, Quality 39, and Process Improvement 34. Figure 1.7 illustrates a network for the 52 words that meet the criterion of being repeated at least seven times, which are integrated into seven clusters, where the first one refers to different tools that are used together with DMAIC, the second one integrates the statistical techniques used, and the third one is a continuous improvement approach. The other four clusters refer to the applications and approaches applied to DMAIC.

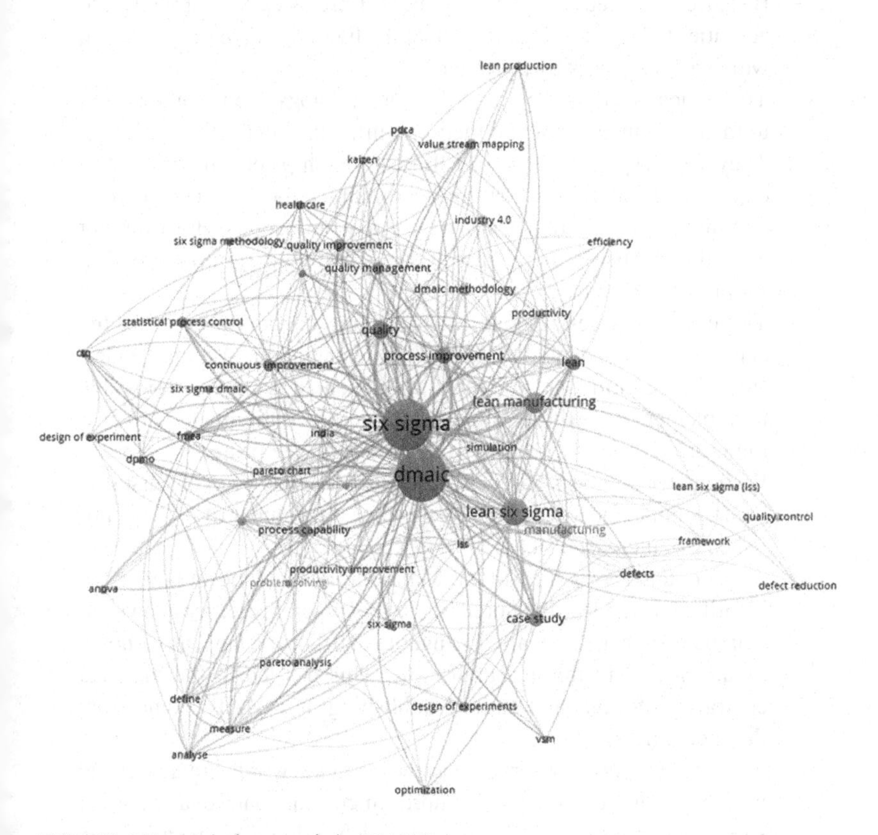

FIGURE 1.7 Main keywords in DMAIC.

1.7 CONCLUSIONS

The concepts, applications, critical success factors, benefits, and barriers of DMAIC, as well as a short bibliometric review, are reviewed in this chapter. The following conclusions were drawn:

- DMAIC is a structured and sequential methodology that serves as the basis for Six Sigma and has yielded economic results in many industrial sectors. It is applied beyond the automotive industry and manufacturing systems, as there are current case studies in the service sector.
- Because of its sequential approach, DMAIC is easy to apply as the activities to be carried out in each of the five stages are fully defined, which avoids errors by beginners.
- The critical success factors of this methodology are mainly related to the top management and their commitments and responsibilities. They are responsible for instilling leadership, eliminating interdepartmental barriers, fostering communication, and establishing continuous improvements as a day-to-day philosophy. However, not everything falls on them since workers and their training are vital for promoting change.
- These same success factors often become barriers preventing the achievement of adequate DMAIC implementation. Without leadership and commitment from top management, it is almost impossible to implement it, as there would not be a correct alignment with customer needs and distribution of resources for the projects.
- Bibliometric analysis has shown that DMAIC has been of outstanding academic and scientific interest since the number of papers published has increased steadily.
- Asian countries such as India, Malaysia, and Indonesia have been identified as the leading countries in this type of research; however, countries such as the United Kingdom and Europe have also shown broad interest in this research topic. This has led to the fact that central institutions and authors publishing on DMAIC come from these countries.
- Finally, it has been observed that the main keywords are associated with Six Sigma, continuous improvement, and lately, a series of word combinations have been found, such as Lean Sigma and Lean Six Sigma.

REFERENCES

Abhilash CR, Thakkar JJ (2019) Application of Six Sigma DMAIC methodology to reduce the defects in a telecommunication cabinet door manufacturing process: A case study. Int J Qual Reliab Manag IJQRM-12-2018-0344. https://doi.org/10.1108/IJQRM-12-2018-0344

Ahmed A, Olsen J, Page J (2023) Integration of Six Sigma and simulations in real production factory to improve performance—a case study analysis. Int J Lean Six Sigma 14:451–482. https://doi.org/10.1108/IJLSS-06-2021-0104/FULL/XML

Aichouni ABE, Abdullah H, Ramlie F (2021) A scientific approach of using the DMAIC methodology to investigate the effect of cutting tool life on product quality and process economics: A case study of a Saudi manufacturing plant. Eng Technol Appl Sci Res 11:6799–6805. https://doi.org/10.48084/ETASR.4008

Albliwi S, Antony J, Lim SAH, et al. (2014) Critical failure factors of Lean Six Sigma: A systematic literature review. Int J Qual Reliab Manag 31:1012–1030. https://doi.org/10.1108/IJQRM-09-2013-0147/FULL/XML

Alhuraish I, Robledo C, Kobi A (2017) A comparative exploration of lean manufacturing and Six Sigma in terms of their critical success factors. J Clean Prod 164:325–337. https://doi.org/10.1016/J.JCLEPRO.2017.06.146

Al-Qatawneh L, Abdallah AAA, Zalloum SSZ (2019) Six Sigma application in healthcare logistics: A framework and a case study. J Healthc Eng 2019:1–12. https://doi.org/10.1155/2019/9691568.

Alvarez-Rozo DC, Ortiz-Rodríguez OO, Rozo-Rojas I (2020) Application of the Six Sigma methodology in the paste preparation area of a company in the ceramic sector. Respuestas 25:125–141. https://doi.org/10.22463/0122820X.2825

Amrina U, Firmansyah H (2019) Analysis of defect and quality improvement for O ring product through applying DMAIC methodology. J Pasti (Penelitian dan Apl Sist dan Tek Ind 13:136–148. https://doi.org/10.22441/PASTI.2019.V13I2.003

Antony J, Banuelas R (2002) Key ingredients for the effective implementation of Six Sigma program. Meas Bus Excell 6:20–27. https://doi.org/10.1108/13683040210451679/FULL/XML

Araman H, Saleh Y (2023) A case study on implementing Lean Six Sigma: DMAIC methodology in aluminum profiles extrusion process. TQM J 35:337–365. https://doi.org/10.1108/TQM-05-2021-0154/FULL/XML

Banawi AA, Besné A, Fonseca D, et al. (2020) A three methods proactive improvement model for buildings construction processes. Sustainability 12:4335. https://doi.org/10.3390/SU12104335

Ben Ruben R, Vinodh S, Asokan P (2017) Implementation of Lean Six Sigma framework with environmental considerations in an Indian automotive component manufacturing firm: A case study. Prod Plan Control 28:1193–1211. https://doi.org/10.1080/09537287.2017.1357215

Cano JA, Gómez RA, Cortés P (2021) ICT validation in logistics processes: Improvement of distribution processes in a goods sector company. Informatics 8:75. https://doi.org/10.3390/INFORMATICS8040075

Cesarelli G, Petrelli R, Ricciardi C, et al. (2021) Reducing the healthcare-associated infections in a rehabilitation hospital under the guidance of Lean Six Sigma and DMAIC. Healthcare 9:1667. https://doi.org/10.3390/HEALTHCARE9121667

Chen M, Lyu J (2009) A Lean Six-Sigma approach to touch panel quality improvement. Prod Plan Control 20:445–454. https://doi.org/10.1080/09537280902946343

Cheng J-L (2017) Implementing Six Sigma via management by objectives strategy: An empirical study in Taiwan. J Bus Manag Sci 5:35–41. https://doi.org/10.12691/jbms-5-2-2

Condé GCP, Oprime PC, Pimenta ML, et al. (2023) Defect reduction using DMAIC and Lean Six Sigma: A case study in a manufacturing car parts supplier. Int J Qual Reliab Manag 40:2184–2204. https://doi.org/10.1108/IJQRM-05-2022-0157/FULL/XML

Dambhare S, Aphale S, Kakade K, et al. (2013) Productivity improvement of a special purpose machine using DMAIC principles: A case study. J Qual Reliab Eng 2013:1–13. https://doi.org/10.1155/2013/752164

Daniyan I, Adeodu A, Mpofu K, et al. (2022) Application of Lean Six Sigma methodology using DMAIC approach for the improvement of bogie assembly process in the railcar industry. Heliyon 8:e09043. https://doi.org/10.1016/J.HELIYON.2022.E09043

Da Rocha ASC, De Araújo MP, Volscham A, et al. (2010) Evidence of healthcare quality improvement in acute myocardial infarction. Arq Bras Cardiol 94:726–729. https://doi.org/10.1590/S0066-782X2010005000047

El Safty SB (2011) Critical success factors of Six-Sigma implementation in automotive industry in Egypt. SAE 2011 World Congr Exhib. https://doi.org/10.4271/2011-01-1270

Erdil NO, Aktas CB, Arani OM (2018) Embedding sustainability in Lean Six Sigma efforts. J Clean Prod 198:520–529. https://doi.org/10.1016/J.JCLEPRO.2018.07.048

Galvão GDA, Evans S, Ferrer PSS, et al. (2022) Circular business model: Breaking down barriers towards sustainable development. Bus Strateg Environ 31:1504–1524. https://doi.org/10.1002/BSE.2966

Gijo EV, Scaria J, Antony J (2011) Application of Six Sigma methodology to reduce defects of a grinding process. Qual Reliab Eng Int 27:1221–1234. https://doi.org/10.1002/qre.1212

Girmanová L, Šolc M, Kliment J, et al. (2017) Application of Six Sigma using DMAIC methodology in the process of product quality control in metallurgical operation. Acta Technol Agric 20:104–109. https://doi.org/10.1515/ata-2017-0020

Godina R, Silva BGR, Espadinha-Cruz P (2021) A DMAIC integrated fuzzy FMEA model: A case study in the automotive industry. Appl Sci 11:3726. https://doi.org/10.3390/APP11083726

Guo W, Jiang P, Xu L, Peng G (2019) Integration of value stream mapping with DMAIC for concurrent Lean-Kaizen: A case study on an air-conditioner assembly line. Adv Mech Eng 11:1–17. https://doi.org/10.1177/1687814019827115

Hamdan A, Hamdan S, Alsyouf I, et al. (2024) Enhancing sustainability performance of universities: A DMAIC approach. Syst Res Behav Sci 41:153–172. https://doi.org/10.1002/SRES.2942

Hardy DL, Kundu S, Latif M (2021) Productivity and process performance in a manual trimming cell exploiting Lean Six Sigma (LSS) DMAIC—a case study in laminated panel production. Int J Qual Reliab Manag 38:1861–1879. https://doi.org/10.1108/IJQRM-07-2020-0242/FULL/XML

Herrera Gómez M, Trinidad Requena A (2002) Planning as a social process. Gestión Y Análisis Políticas Públicas 61–77. https://doi.org/10.24965/gapp.vi25.335

Hors C, Goldberg AC, Pereira de Almeida EH, et al. (2012) Application of the enterprise management tools Lean Six Sigma and PMBOK in developing a program of research management. Heal Econ Manag 10:480–490. https://doi.org/10.1590/S1679-45082012000400015

Jain D, Dash MK, Thakur KS (2022) Development of research agenda on demonetization based on bibliometric visualization. Int J Emerg Mark 17:2584–2604. https://doi.org/10.1108/IJOEM-12-2019-1085/FULL/XML

Jirasukprasert P, Garza-Reyes JA, Kumar V, et al. (2015) A Six Sigma and DMAIC application for the reduction of defects in a rubber gloves manufacturing process. Int J Lean Six Sigma 5:2–22. https://doi.org/10.1108/IJLSS-03-2013-0020/FULL/XML

Karakose T, Yirci R, Papadakis S, et al. (2021) Science mapping of the global knowledge base on management, leadership, and administration related to COVID-19 for promoting the sustainability of scientific research. Sustainability 13:9631. https://doi.org/10.3390/SU13179631

Khan MA, Ali MK, Sajid M (2022) Lean implementation framework: A case of performance improvement of casting process. IEEE Access 10:81281–81295. https://doi.org/10.1109/ACCESS.2022.3194064

Khan SA, Kaviani MA, Galli B, et al. (2019) Application of continuous improvement techniques to improve organization performance: A case study. Int J Lean Six Sigma 10:542–565. https://doi.org/10.1108/IJLSS-05-2017-0048/FULL/XML

Knop K (2019) Analysis and improvement of the galvanized wire production process with the use of DMAIC cycle. Qual Prod Improv 1:551–558. https://doi.org/10.2478/CQPI-2019-0074

Kumar M, Antony J, Singh RK, et al. (2006) Implementing the Lean Sigma framework in an Indian SME: A case study. Prod Plan Control 17:407–423. https://doi.org/10.1080/09537280500483350

Kumar P, Singh D, Bhamu J (2021) Development and validation of DMAIC based framework for process improvement: A case study of Indian manufacturing organization. Int J Qual Reliab Manag 38:1964–1991. https://doi.org/10.1108/IJQRM-10-2020-0332/FULL/XML

Kumar Sharma R, Gopal Sharma R (2014) Integrating Six Sigma culture and TPM framework to improve manufacturing performance in SMEs. Qual Reliab Eng Int 30:745–765. https://doi.org/10.1002/QRE.1525

Li P, Jiang P, Zhang G (2019) An enhanced DMAIC method for feature-driven continuous quality improvement for multi-stage machining processes in one-of-a-kind and small-batch production. IEEE Access 7:32492–32503. https://doi.org/10.1109/ACCESS.2019.2900461

Maciel-Monteon M, Limon-Romero J, Gastelum-Acosta C, et al. (2020) Measuring critical success factors for Six Sigma in higher education institutions: Development and validation of a surveying instrument. IEEE Access 8:1813–1823. https://doi.org/10.1109/ACCESS.2019.2962521

Nagi A, Altarazi S (2017) Integration of value stream map and strategic layout planning into DMAIC approach to improve carpeting process. J Ind Eng Manag 10:74–97. https://doi.org/10.3926/jiem.2040

Näslund D (2013) Lean and six sigma—critical success factors revisited. Int J Qual Serv Sci 5:86–100. https://doi.org/10.1108/17566691311316266/FULL/XML
Navarro-Romero EDC, Mendoza-Moreno JF, García-Corrales N, et al. (2022) Proposal for the integration of risks in the life cycle of business processes and their relationship with process mining. DYNA 89:150–158. https://doi.org/10.15446/dyna.v89n223.102252
Neuscheler-Fritsch D, Norris R (2001) Capturing financial benefits from six sigma. Qual Prog 34:39–44.
Niñerola A, Ferrer-Rullan R, Vidal-Suñé A (2020) Climate change mitigation: Application of management production philosophies for energy saving in industrial processes. Sustainability 12:717. https://doi.org/10.3390/SU12020717
Noori B, Latifi M (2018) Development of Six Sigma methodology to improve grinding processes: A change management approach. Int J Lean Six Sigma 9:50–63. https://doi.org/10.1108/IJLSS-11-2016-0074/FULL/XML
Panayiotou NA, Stergiou KE (2023) Development of a retail supply chain process reference model incorporating Lean Six Sigma initiatives. Int J Lean Six Sigma 14:209–251. https://doi.org/10.1108/IJLSS-04-2021-0079/FULL/XML
Pavletić D, Soković M (2002) Six Sigma: A complex quality initiative. Stroj Vestnik/J Mech Eng 48:158–168.
Powell D, Lundeby S, Chabada L, et al. (2017) Lean Six Sigma and environmental sustainability: The case of a Norwegian dairy producer. Int J Lean Six Sigma 8:53–64. https://doi.org/10.1108/IJLSS-06-2015-0024/FULL/XML
Prashar A (2014) Adoption of Six Sigma DMAIC to reduce cost of poor quality. Int J Product Perform Manag 63:103–126. https://doi.org/10.1108/IJPPM-01-2013-0018/FULL/XML
Rodriguez RD, Medini K, Wuest T (2022) A DMAIC framework to improve quality and sustainability in additive manufacturing-a case study. Sustainability 14:581. https://doi.org/10.3390/SU14010581
Samanta M, Virmani N, Singh RK, et al. (2024) Analysis of critical success factors for successful integration of Lean Six Sigma and Industry 4.0 for organizational excellence. TQM J 36:208–243. https://doi.org/10.1108/TQM-07-2022-0215/FULL/XML
Scaff AV (2002) Six Sigma: Behind the scenes. SAE Tech Pap 1:3510. https://doi.org/10.4271/2002-01-3510
Scheller AC, Sousa-Zomer TT, Cauchick-Miguel PA (2021) Lean Six Sigma in developing countries: Evidence from a large Brazilian manufacturing firm. Int J Lean Six Sigma 12:3–22. https://doi.org/10.1108/IJLSS-09-2016-0047/FULL/XML
Socconini L (2023) The Lean Six Sigma DMAIC methodology explained. https://leansixsigmainstitute.org/the-lean-six-sigma-dmaic-methodology-explained/. Accessed 4 December 2023.
Souza DL, Korzenowski AL, Alvarado MM, et al. (2021) A systematic review on lean applications' in emergency departments. Healthcare 9:763. https://doi.org/10.3390/HEALTHCARE9060763
Stankalla R, Koval O, Chromjakova F (2018) A review of critical success factors for the successful implementation of Lean Six Sigma and Six Sigma in manufacturing small and medium sized enterprises. Qual Eng 30:453–468. https://doi.org/10.1080/08982112.2018.1448933
Subagyo IE, Saraswati D, Trilaksono T, et al. (2020) Benefits and challenges of DMAIC methodology implementation in service companies: An exploratory study. J Apl Manaj 18:814–824. https://doi.org/10.21776/UB.JAM.2020.018.04.19

Swarnakar V, Singh AR, Antony J, et al. (2020) A multiple integrated approach for modelling critical success factors in sustainable LSS implementation. Comput Ind Eng 150:106865. https://doi.org/10.1016/J.CIE.2020.106865

Syaifoelida F, Ying CP (2020) The productivity performance's measurement in SME industry by using DMAIC of Six Sigma. Int J Innov Technol Explor Eng 9:2708–2713. https://doi.org/10.35940/IJITEE.C9230.019320

Thomas AJ, Francis M, Fisher R, et al. (2016) Implementing Lean Six Sigma to overcome the production challenges in an aerospace company. Prod Plan Control 27:591–603. https://doi.org/10.1080/09537287.2016.1165300

Trubetskaya A, Ryan A, Powell DJ, et al. (2023) Utilising a hybrid DMAIC/TAM model to optimise annual maintenance shutdown performance in the dairy industry: A case study. Int J Lean Six Sigma 15:70–92. https://doi.org/10.1108/IJLSS-05-2023-0083/FULL/PDF

Vasconcellos de Araujo B (2020) Lean Six Sigma in services: An application of the methodology in the attendiment sector of an exam laboratory. Sci J Bus Manag 8:119–131. https://doi.org/10.11648/J.SJBM.20200803.13

Velu SJS, Jusoh MS, Muhd Yusuf DH, et al. (2021) Six Sigma in semiconductor: Continuous improvement in production floor area. J Phys Conf Ser 2129:012036. https://doi.org/10.1088/1742-6596/2129/1/012036

Vinodh S, Kumar SV, Vimal KEK (2014) Implementing Lean Sigma in an Indian rotary switches manufacturing organization. Prod Plan Control 25:288–302. https://doi.org/10.1080/09537287.2012.684726

Wang FK, Chen KS (2010) Applying Lean Six Sigma and TRIZ methodology in banking services. Total Qual Manag 21:301–315. https://doi.org/10.1080/14783360903553248

Wibowo R (2019) Impact of procurement professionalization on the efficiency of public procurement. Oper Excell J Appl Ind Eng 11:228–239. https://doi.org/10.22441/OE.V11.3.2019.032

Woodall WH, Driscoll AR, Montgomery DC (2022) A review and perspective on neutrosophic statistical process monitoring methods. IEEE Access 10:100456–100462. https://doi.org/10.1109/ACCESS.2022.3207188

Yadav N, Mathiyazhagan K, Kumar K (2019) Application of Six Sigma to minimize the defects in glass manufacturing industry: A case study. J Adv Manag Res 16:594–624. https://doi.org/10.1108/JAMR-11-2018-0102/FULL/XML

Yuan W, Li Z, Han J, et al. (2023) Improving the resident assessment process: Application of App-based e-training platform and lean thinking. BMC Med Educ 23:1–9. https://doi.org/10.1186/S12909-023-04118-2/TABLES/3

Zakari MA (2014) Challenges of International Financial Reporting Standards (IFRS) adoption in Libya. Int J Account Financ Report 4:390–412. https://doi.org/10.5296/IJAFR.V4I2.6302

Zdęba-Mozoła A, Kozłowski R, Rybarczyk-Szwajkowska A, et al. (2023) Implementation of lean management tools using an example of analysis of prolonged stays of patients in a multi-specialist hospital in Poland. Int J Environ Res Public Health 20:1067. https://doi.org/10.3390/IJERPH20021067/S1

Auxiliary Tools in DMAIC

2

2.1 INTRODUCTION

The DMAIC methodology is structured; however, specific tools and techniques can be used within the five phases (Define, Measure, Analyze, Improve, Control). The following is a list of these tools and techniques, which does not indicate that they are the only ones but the most common or indispensable.

(1) Definition phase
- Suppliers, Inputs, Process, Outputs, and Customers (SIPOC): This tool helps to define process boundaries and stakeholders.
- Voice of the Customer (VOC): Techniques to capture customer needs and expectations.
- Stakeholder analysis: This technique is used to identify stakeholders and their needs.
- Project charter: This document describes the objectives, scope, schedule, and resources of the project.
- Process flow diagrams: Allows knowledge of the sequence in which the activities are carried out.

(2) Measurement phase
- Data collection techniques: Surveys, interviews, observations, and control charts
- Process mapping: Visual representation of the process steps.
- Key Performance Indicators (KPIs): Definition of metrics for tracking performance.

(3) Analysis Phase
- Cause and effect diagrams (Ishikawa Fishbone): Identification of possible causes of defects.

DOI: 10.1201/9781003564607-2

- Pareto diagrams: Prioritization of problems according to frequency or impact.
- Statistical analysis: Tools such as hypothesis testing to analyze quantitative data.
- Analysis of variance (ANOVA): Allows the identification of variations in processes.

(4) Improvement Phase
- Brainstorming: Creating creative solutions to address root causes.
- Failure Mode and Effect Analysis (FMEA): Identification of potential failure points and mitigation strategies.
- Design of experiments: Allows the testing of solutions and identifies the best solution.
- Pilot programs: Testing a solution on a small scale before full implementation.

(5) Control Phase
- Control charts: Monitoring process performance after implementing changes.
- Standard Operating Procedures (SOP): Documentation of improved process.
- Poka-Yoke: Error-proof mechanisms to prevent defects.

This book does not intend to define each technique and tool; however, the most important ones are defined in what follows.

2.2 TOOLS AND TECHNIQUES IN THE DEFINITION PHASE

2.2.1 SIPOC

The Suppliers, Inputs, Process, Outputs, Outputs, and Customers (SIPOC) tool is used in process management to clearly and concisely understand a process's scope, essential elements, and expected outcomes (Zaman and Zerin 2017). This tool defines and documents the following elements (Silvani et al. 2019).

- Suppliers (S): Identify the sources (people or companies) of information, materials, or resources that feed the process.
- Inputs (I): The inputs or resources needed to initiate the process can be data, materials, information, and time.

- Process (P): Describes the activities and steps in transforming inputs into outputs.
- Outputs (O): This process generates expected results or products.
- Customers (C): Persons, departments, or entities that receive and use the outputs of the process.

SIPOC allows the inputs and outputs of a process to be visualized in a tabular manner (Goswami 2018) and to identify and understand the relationships between a system's inputs, processes, and outputs. Presenting a summary from the input to the process output in tabular form facilitates understanding all the stages involved (González-González and Escobar-Prado 2021).

This tool provides a clear view of the input sources, process, and outputs and helps identify the customers involved (Bertocci et al. 2021). Therefore, SIPOC can be used to balance the flow of information because all the information received comes from a specific function of another system (Kartinawati et al. 2022).

Applying the SIPOC diagram involves mapping and identifying the suppliers, inputs, processes, outputs, and customers involved in a process (Farrukh et al. 2020). This tool is widely used in the project definition phase and is fundamental for identifying processes that require improvement (Kartinawati et al. 2022). In addition, developing an SIPOC diagram based on the key inputs from the Voice of the Customer clarifies the process that needs improvement (Subagyo et al. 2020).

The advantages of using a SIPOC diagram in a company are as follows (QuestionPro 2023):

(1) Clarity in understanding the process: The SIPOC diagram provides a clear and concise view of the process, which helps team members and stakeholders understand how the process is carried out and how different elements are related.
(2) Identification of critical elements: The SIPOC diagram helps to identify the critical elements of the process, allowing teams to focus their efforts on the most important areas for process improvement.
(3) Identification of improvement opportunities: By identifying the critical elements of the process, the SIPOC diagram helps to identify opportunities for improvement and establish clear objectives for achieving process improvements.
(4) Effective communication: The SIPOC diagram is an effective communication tool that allows team members and stakeholders to understand the process and its relationships with different elements quickly.

On the other hand, the disadvantages of using a SIPOC diagram in companies are (QuestionPro 2023):

(1) Oversimplification: The SIPOC diagram can oversimplify the process's reality, leading to incomplete or superficial understanding.
(2) Lack of detail: The SIPOC diagram does not provide details on how activities are carried out within the process, making it challenging to identify specific issues and troubleshoot problems.
(3) Lack of focus on quality: The SIPOC diagram focuses on the sequence of the process but not on the quality of the process results.

2.2.2 Voice of the Customer

The voice of the customer (VOC) is a critical component in the implementation of the DMAIC methodology as it encapsulates the needs, expectations, and preferences of customers, providing valuable information that directs efforts toward process improvement within companies. Therefore, understanding VOC is vital for defining project objectives, identifying quality-critical characteristics, and ensuring that process improvements conform to customer requirements (Cano et al. 2020).

In the DMAIC methodology, VOC is typically acquired during the Define phase, where project objectives are established and customer requirements are clearly outlined (Gupta et al. 2018; Hardy et al. 2021). This phase involves surveys, interviews, focus groups, and data analysis to prioritize requirements (Tsarouhas and Sidiropoulou 2024; Condé et al. 2023). By integrating OCV from the outset, organizations can focus on identifying suitable projects to refine processes to meet customer expectations and increase customer satisfaction and loyalty.

Thus, the VOC is the cornerstone of the entire DMAIC process during the measurement phase. Data were collected to quantify the current performance of the process in meeting the customer requirements identified in the definition phase (Subagyo et al. 2020; Prashar 2014). Analysis of this data in the subsequent Analyze phase enables organizations to identify the root causes of problems affecting customer satisfaction (Prashar 2014; Noori and Latifi 2018), and by correlating process performance metrics with customer needs, opportunities for improvement are prioritized.

In the improvement phase, solutions are devised and implemented based on insights gained from the analysis of VOC and process data (Noori and Latifi 2018; Condé et al. 2023). Continuous feedback from customers is crucial to validate the effectiveness of solutions and ensure that they match expectations (Prashar 2014). Finally, in the Control phase, mechanisms are put in place to sustain the improvements made and monitor ongoing performance to ensure that customer requirements continue to be met (Prashar 2014; Condé et al. 2023)

Therefore, it was concluded that VOC offers several advantages in its application, which can be summarized as follows (Condé et al. 2023; Prashar 2020; Adeodu et al. 2020):

(1) Customer-centric approach
(2) Improved product quality
(3) Cost savings
(4) Competitive advantage
(5) Continuous improvement
(6) Employee engagement
(7) Data-driven decision making
(8) Sustainable business growth

However, using VOCs also has disadvantages; therefore, it is essential to remember that much of the information is obtained subjectively through surveys or that customer preferences change from one day to the next (Tsarouhas and Sidiropoulou 2024). Other authors indicate that VOC is very limited in scope, as it does not consider market trends or generate inconsistent comments from someone considered an expert. In addition, addressing the needs and voices of the customer requires making changes within the firm, which generally offers resistance (Araman and Saleh 2023).

In addition, it is reported that there is often an excessive expenditure of resources in trying to understand customer needs, which can be time and money (Subagyo et al. 2020). Other authors report that customers often complain that the manufacturer demands too much feedback on some aspects, which is often tiring for the survey respondents or meeting attendees (Patyal and Koilakuntla 2016). Likewise, it has been found that it is often difficult to measure the impact of actions within the company and their repercussions on customer satisfaction, which are often subjective and unrealistic (Cano et al. 2021).

Finally, it is important to mention that most of the information generated at this stage is not standardized (Baro et al. 2023). Therefore, it may have the risk of being misinterpreted, which would cause companies to execute or carry out projects that do not meet the customer's real needs (Vendrame Takao et al. 2017).

2.2.3 Stakeholder Analysis

Stakeholders are interested parties, individuals, or groups interested in a project or affected by their results. Therefore, it is vital to integrate them into

DMAIC projects. The importance of these results can be summarized as follows (Sunder M 2016; Knezek et al. 2022):

(1) Identification of stakeholders, which may be internal and associated with employees and the administrative part of the company or with the external part that relates to customers, such as suppliers, regulatory or governmental bodies, and the community in general.
(2) Understanding the needs and expectations of stakeholders is vital to aligning project objectives with those of all stakeholders and seeking a balance between them.
(3) Stakeholder influence management is essential for understanding stakeholder power dynamics, as organizations must effectively manage the relationships and conflicts between stakeholders.
(4) Improving communication and engagement by involving stakeholders in solving a problem fosters internal and external collaboration, builds trust, and ensures that their opinions are integrated into the decision-making process.
(5) Mitigation of risks and challenges, since the joint implementation of activities implies a shared risk to the different challenges that may be faced, may reduce resistance to change in projects.
(6) When problems or challenges are identified, all stakeholders' participation helps generate collaborative strategies that allow easy adaptation to meet the needs of all participants, which facilitates the decision-making process promptly.
(7) Finally, it is essential to mention that involving all stakeholders facilitates monitoring and adaptation because it is easier for a member to identify deviations promptly and take corrective actions.

Stakeholder analysis, when implementing the DMAIC methodology, offers many benefits widely sought by managers, as it allows for understanding involving all stakeholders and fostering collaboration towards continuous improvement and quick decision-making. A summary of these benefits is provided in what follows:

(1) Improved decision-making by integrating all parties' expectations facilitates aligned decision-making on a common goal (Kaur and Lodhia 2018).
(2) Improved stakeholder involvement, as there is an increased sense of ownership and belonging, generates greater stakeholder commitment. This, in turn, allows the building of stronger relationships among all constituents, rapidly fostering direct communication (Bellucci et al. 2019).

(3) Increased transparency of the developed processes and activities because they are known to all. This gives credibility and generates trust in the decision-making process, which fosters collaboration among all the stakeholders (Ubaid and Dweiri 2024)
(4) Mitigating factors that may jeopardize the success of projects, as all stakeholders are involved, follow up on activities, quickly become aware of deviations, and promptly take corrective actions (Trubetskaya et al. 2024).
(5) Alignment with the needs of the stakeholders in the different meetings allows prioritizing the projects to be implemented according to the needs of the whole group (Panayiotou et al. 2022).
(6) Improved and rapid innovation to provide solutions because, as a group, the experiences and knowledge of many people are integrated, which must be used to the maximum (Gagné et al. 2022).
(7) Quality improvements in processes and products are identified as critical areas for improvement, and corrective activities or actions are prioritized to solve them. This allows for the identification of deviations at an early stage (Daudelin et al. 2020).
(8) Sustainable relationships among all group members, as effective and long-term communications are fostered, undoubtedly lead to the cultivation of sustainable and ongoing collaborative relationships (Martínez et al. 2021; Sarwar et al. 2022).
(9) Finally, companies gain a competitive advantage by leveraging parties' perspectives and focusing on solving quality problems, often allowing them to differentiate themselves in the marketplace and enhance their reputation (Martínez et al. 2021).

However, it is important to highlight that several authors have mentioned a series of disadvantages when working with all stakeholders, including great difficulty in managing and controlling the interests of the whole group or resistance to change encountered by some of those involved due to conflicts of interest. (Swarnakar and Vinodh 2016). Likewise, it has been reported that some of the stakeholders often do not show significant participation or involvement in the solution of the problems, and they are usually the ones who complain about the results obtained at the end of the project (Jamil et al. 2020).

Other authors mention that stakeholders in globalized industries may find communication difficult because of language limitations. Not all of them have the resources to invest in solving the problem, often leading to inconsistent project prioritization (Prashar 2020).

Finally, access is often required for information that may be confidential and cannot be shared with everyone else because it is part of trade secrets or company strategies (Hamdan et al. 2024). In addition, it is often challenging to measure stakeholders' impact, and it is difficult to determine whether they should be part of the group, as sometimes they are more of a problem than support (Tufail et al. 2022).

2.2.4 Project Charter

The project charter is a formal document that defines the scope, objectives, schedule, budget, and resources of a DMAIC project, making it a tool to ensure that the project is implemented effectively and efficiently but provides a quick summary to any of the stakeholders (Trakulsunti and Antony 2018). The project charter is usually developed in the Define stage of the DMAIC methodology, and its content may vary depending on the organization and complexity of the project; however, it should include the name of the project, in brief, a description of the project stating the purpose and scope; the project objectives, which should be specific, measurable, achievable, relevant, and time-bound (SMART); the scope of the project, indicating what is and is not included within the scope; a project timeline outlining key dates, start dates, completion dates, and critical milestones related to review meetings; and a project timeline that describes the project's objectives (Jamil et al. 2020), project budget indicating estimated costs, project team indicating the names and roles of the people involved, identified risks and mitigation proposals, and approvals with signatures giving the required formality (Ubaid and Dweiri 2024).

Adding a project charter at the beginning of the project provides a quick summary of what you want to do with the project, which provides clarity and helps align all stakeholders toward a common goal. Everyone can have a general idea of what they want to do. In addition, such a charter helps to identify opportunities quickly, as all efforts are focused on operations that are vital to the project or have the most significant impact (Trakulsunti and Antony 2018).

In addition, the project charter facilitates communication and collaboration among team members, stakeholders, and management by establishing a shared understanding of the project's objectives and expectations (Laureani and Antony 2012). It also aids in risk management by identifying uncertainties at the beginning of the project, which generates proactive actions (Marques and Matthé 2017).

However, some authors also point out that their use may have disadvantages associated with a lack of clarity and specificity, as they summarize the entire intent of the project, and some team members may be the only thing they read (Ahammed and Hasan 2020). This can lead to a lack of understanding of expected goals and possible outcomes. In addition, such a summary may not integrate the actual nature with which the project is intended to be executed, especially in companies where production processes and organizational structures are complex and frequently change (Patyal et al. 2021).

2.2.5 Process Flow Diagram

In the manufacturing industry, a process flow diagram is a crucial tool for understanding and analyzing the workflow and steps involved in producing a product or service (Li et al. 2021). This tool provides a visual representation of all stages of the process from start to finish, showing the sequence of activities and how they relate to each other (Barosz et al. 2020). The process flow diagram identifies possible areas of improvement, bottlenecks, or inefficiencies in the process, which help optimize production and ensure an efficient workflow (NSB et al. 2022). In addition, this diagram can also help effectively communicate the process to all stakeholders, which is especially useful when dealing with interdepartmental collaboration or when there are changes in personnel or roles (Cazacu et al. 2021). Diagram development begins with a review stage, where data are received and necessary checks are made. This is followed by the analysis stage, in which the data are examined in more detail, and decisions are made based on the results. Finally, it concludes with the reporting stage, where a final report is produced, and the results are communicated to stakeholders (Sulthan et al. 2021).

Using the process flow diagram, organizations can identify opportunities for improvement in their production chain and take steps to optimize efficiency and productivity (Barosz et al. 2020). In addition, the process flow diagram can also be used to identify areas where just-in-time manufacturing practices can be implemented, such as the Toyota plant in Japan. This reduces energy consumption and waste production, thus benefiting companies and the environment. The process flow diagram can also identify areas where efficient and sustainable manufacturing practices can be implemented (Barosz et al. 2020). This includes implementing supply chain management processes that optimize processes and reduce production and delivery cycles, enabling companies to adapt to market changes and meet customer demands more quickly and efficiently. Using a process flow diagram to manage organizations in business and healthcare has become a critical tool for improving operational efficiency and promoting innovation and compliance with products or services (Omair et al. 2017).

In summary, the process flow diagram is essential for understanding and optimizing manufacturing processes. It provides a clear visual representation of the activities and their sequence.

The process flow diagram offers the following advantages:

(1) It visualizes the stages and sequences of a process (Hernández-Soto et al. 2021a).
(2) This facilitates the identification of possible bottlenecks or points for process improvement (Sulthan et al. 2021).
(3) It helps identify energy and material flows in a process, improving energy efficiency and reducing resource consumption (Santosa and Mulyana 2023).

(4) This facilitates identifying opportunities to reduce energy consumption and waste production (Santosa and Mulyana 2023).
(5) This helps to understand better the added value of each stage of the process and identify possible improvement areas in efficiency and quality (Duc et al. 2022).
(6) It helps identify and eliminate unnecessary activities, increases productivity, and reduces cycle times.

However, the process flow diagram has some disadvantages, including the following:

(1) Interpretation can be complex, especially for those unfamiliar with the diagram's notation (Santosa and Mulyana 2023).
(2) This can become too detailed and extensive, making it difficult to understand and follow (Santosa and Mulyana 2023).
(3) It may not capture all interactions and variables in a given process, which can lead to incorrect decisions or inaccuracies in process analysis (Valier 2020).
(4) It can quickly become obsolete, especially in constantly changing environments where processes can evolve and change frequently (Valier 2020).

2.3 TOOLS AND TECHNIQUES IN THE MEASUREMENT PHASE

2.3.1 Data Collection Techniques

Many techniques are used to gather information and make appropriate decisions when implementing DMAIC methodology in the industry; therefore, only a few are mentioned in what follows. Readers exploring concepts, definitions, and procedures should read more specific books.

- Measurement system analysis (MSA): This technique is part of the comprehensive strategy for collecting critical-to-quality (CTQ) parameters and includes the type of data, unit of measurement, sampling methods, and accuracy of the measurement system (Tsarouhas and Sidiropoulou 2024).
- Design of Experiments (DOE): This is based on hypothesis testing and analysis of variance techniques to determine process variables'

statistical impact on defects and identify optimal values for process improvement (Jirasukprasert et al. 2015).

- Regression analysis and hypothesis testing: This technique allows the validation of relationships between variables through hypothesis testing, the Taguchi method, classification techniques, and regression trees (Antony et al. 2012).
- Data recording and establishment of the baseline: This allows for identifying the data's characteristics, studying the measurement system's accuracy, and recording the data (Gijo et al. 2014).
- Seven essential quality tools: These methodologies help to reduce defects, improve process performance, and identify possible relationships between quality and process performance (Aichouni et al. 2021).
- Lean and Six Sigma principles: These allow structuring the collection and analysis of data under a continuous improvement-based approach (Powell et al. 2017).
- Multi-criteria decision-making methods (MCDM): This helps in decision-making when there is more than one parameter to be considered in the process of identifying the idealized alternative (Trubetskaya et al. 2023)
- Semi-structured interviews, observations, and web questionnaires were used as customer data collection methods (Johansson et al. 2020).

Some authors have investigated the advantages of the data collection techniques used when implementing DMAIC, among which are the high statistical rigor with which the tests are conducted (Flynn et al. 1990) and the continuous monitoring of all vital variables. In addition, they make it possible to quickly find optimum production levels through structured techniques (Tsarouhas and Sidiropoulou 2024). This facilitates real-time decision-making and continuous improvement.

However, other authors mention that this information may have some statistical bias, with too much or too little variance, leading to the suspicion that the process information often lacks the required quality or is unreliable (Clancy et al. 2023). Likewise, it has been reported that these statistical techniques are often complex and not easily understood by all work team members, so very superfluous analyses can be made that lack the required depth (Johansson et al. 2020).

2.3.2 Process Mapping

Process mapping is a popular Six Sigma framework for process improvement that is used in the Define phase of the DMAIC methodology. In essence, it acts as a visual blueprint for a process that helps identify areas for improvement.

That is, it visually illustrates the steps of a process to detect inefficiencies, redundancies, and areas of improvement. This allows organizations to understand how activities are interconnected and where bottlenecks or errors may arise that could affect the process (Indra et al. 2021).

Therefore, it is concluded that process mapping focuses on a better understanding of the process, which is shared among team members about how activities are carried out. It describes the sequence of steps or activities, inputs, outputs, decision points, and those involved at each stage, facilitating the direction and execution of improvement plans (Morlock and Meier 2015). Also, process mapping seeks to identify inefficiencies, which allows for easy detection of bottlenecks, delays, and activities that do not add value but cost the customer to pay (Rimantho and Sari 2023)

Process mapping uses a set of components, and the first is standard symbols to indicate different elements, such as tasks (rectangles), decisions (diamonds), documents (pages), and data flows (arrows) (Soltani et al. 2020). The second component is the steps or sequences, where each step involved in the process is clearly defined and the order in which they should be executed. The third component comprises the inputs and outputs representing the raw materials, information, or data entering the process (inputs) and the final product or result (outputs). However, in a process sequence, it is possible to have decision points where it bifurcates and an alternative must be chosen (Lorenzon dos Santos et al. 2019). Finally, the fifth common indicates the people involved and the individuals or departments responsible for completing each step within the sequence.

Therefore, mapping production processes offers many advantages, the most important of which are as follows.

(1) Identify process waste such as excess production, waiting times, unnecessary transportation, inventory, defects, and underutilized talent (Salwin et al. 2021).
(2) Process optimization involves gaining a comprehensive end-to-end understanding of these processes. This allows for the identification of bottlenecks, redundancies, and areas for improvement, thereby facilitating the implementation of lean principles (Shamsu Anuar and Mansor 2022).
(3) It improved productivity by enabling the quantification of waste, labor productivity, quality, and production time.
(4) Customer satisfaction is achieved by increasing product quality, reducing costs, and ensuring timely delivery (Jasti et al. 2020).
(5) Sustainability by identifying polluting sources, which allows the generation of environmentally responsible production processes (Budihardjo and Hadipuro 2022).

However, other authors have mentioned that it is possible that when applying process mapping, a series of problems are encountered, such as the complexity of the processes that make their graphical representation difficult, which often forces them to synthesize it (De Steur et al. 2016). In addition, it has been mentioned that it can be subjective and that stakeholders may understand the same process or activity differently (Jasti et al. 2020). In addition, it has been reported that because of these differences between the parties involved, there may be resistance to change when implementing improvements, or that information is often inaccurate, and there is little flexibility to make changes (Abdel-Jaber et al. 2022).

2.3.3 Key Performance Indicators (KPIs)

Key performance indicators (KPIs) are metrics used in various industries to measure the success and effectiveness of processes, projects, and organizations by providing performance information and enabling informed decision-making and continuous improvement. They usually monitor progress toward organizational goals and objectives (Tambare et al. 2022). Depending on the industry or service sector, they are the KPIs that companies should have; for example, they have been reported in the construction sector (Nourbakhsh et al. 2012), supply chain management (Hatmoko and Neilkelvin 2023), industrial systems (Sun et al. 2023), mining industry (Gackowiec et al. 2020; Mutingi et al. 2016), and hospitality (Widz et al. 2022; Dasandara et al. 2022).

KPIs can be objective or subjective and have been developed based on a comprehensive literature review (Moktadir et al. 2020). They have been reported to be project success (Augustínová and Daubner 2014), productivity (Bilen et al. 2022; Lastochkina 2021), quality management (Petkova et al. 2023) and sustainability (Moktadir et al. 2020). However, KPIs are also used to ensure transparency, measurability, and optimal management decision-making in companies (Docekalová et al. 2018).

The benefits of using KPIs in the industry can be summarized as follows:

(1) They allow a structured follow-up of the goals and objectives that the company intends to improve, guaranteeing their alignment with pre-established strategies (Maté et al. 2017).
(2) They facilitate improved decision-making based on updated information because data are obtained constantly and periodically, which allows corrective actions to be taken when deviations occur (Maté et al. 2017).
(3) Improve competitiveness and performance because areas of opportunity are quickly identified as well as strengths, which allow the exploitation of resources to improve competitiveness (Rodrigues et al. 2021).

(4) They allow for improved operational efficiency as they are the first indices to be monitored and tracked consistently, leading to greater operational efficiency and effectiveness (Rodrigues et al. 2021).
(5) They help with environmental sustainability and responsibility, as environmental aspects are monitored, evaluated, and reported to governmental entities. Such monitoring and control avoid economic and administrative penalties and encourage sustainable practices (Muhammad et al. 2018).

However, there are also several disadvantages to using KPIs, including the following:

(1) There are often limitations in the accuracy and neutrality of KPIs, resulting in biased or inaccurate assessments of performance (Krasodomska and Zarzycka 2021).
(2) In many industrial sectors, there is a strong focus on short-term results, which affects strategic and long-term goals and hinders the ability to focus on sustainable growth and development (Li et al. 2020).
(3) The selection and use of KPIs are unbalanced, leading to a biased view of performance that does not consider all the critical aspects of the organization (Bilen et al. 2022). This is because of the monitoring of incorrect metrics.
(4) Complexity and challenges in developing appropriate KPIs are due to organizational complexity or lack of incentives, top management support, or a culture unfavorable to performance measurement focused on continuous improvement (Gackowiec et al. 2020).
(5) Misleading signals because traditional measures of performance, such as occupancy rates and return on investment, can provide poor and misleading signals that reflect the need for change (Krasodomska and Zarzycka 2021).

2.4 TOOLS AND TECHNIQUES IN THE ANALYSIS PHASE

2.4.1 Cause and Effect Diagram

A cause-and-effect diagram, also known as an Ishikawa diagram or fishbone diagram, is a graphical tool used in quality management in manufacturing industries (Hekmatpanah 2011). This diagram visually represents the potential causes contributing to a specific problem or effect, with the effect being the main problem or result at the head of the "fishbone" and the causes branching out like the bones of the fishbone (E. Ilori et al. 2020). The cause-effect diagram helps to identify, classify, and show the potential causes of a problem, which makes it a valuable problem-solving problem (Sucipto et al. 2022).

The main components of a cause-and-effect diagram typically include the following (Sucipto et al. 2022):

(1) Effect: This refers to the main problem or result being analyzed and is represented at the head of the diagram; in other words, it is the central problem to which the causes are linked.
(2) Causes: These are the factors or reasons that contribute to the effect being analyzed. The causes are usually classified as branches, starting from the main line, representing fish bones.
(3) Categories of causes: Causes are usually grouped into categories to facilitate analysis, and the most commonly used are 4 M (Labor, Machinery, Methods, Materials) or 6 M (adding Measurement and Mother Nature).
(4) Sub-causes: Each category can be hierarchically broken down into sub-causes, leading to the identification of more profound causes for the problem.
(5) Analysis: Because the diagram visually represents cause-effect relationships, it allows for a systematic analysis of the factors contributing to the main problem.

Many uses of cause-effect diagrams have been reported in industry; for example, in the petroleum industry, they have been applied to processes such as oil canning to identify and address potential problems (Hekmatpanah 2011). In the automotive industry, they have been used to reduce the number of defects in paint shops (Memon et al. 2019). Furthermore, in the cocoa agroindustry, cause-effect diagrams are part of the application of statistical process control to improve quality (Sucipto et al. 2022).

From the previous, a series of benefits or advantages of using cause-effect diagrams in the industry can be mentioned, among which the following are indicated:

(1) The cause-effect diagram provides a structured approach to problem-solving by visually mapping the potential causes of a specific problem, facilitating a systematic analysis of the problem (Liliana 2016).
(2) Helps identify the root causes of problems rather than merely address surface symptoms, leading to more effective, sustainable, and reality-representing solutions (Reilly et al. 2014).

(3) The visual representation of relationships in the cause-effect diagram allows informed decisions based on a comprehensive understanding of the factors influencing a particular problem (Gartlehner et al. 2017).
(4) The cause-effect diagram is a communication tool that stakeholders can easily understand, promoting effective communication and collaborative problem-solving (Hamid et al. 2017).
(5) The cause-effect diagram leads to quality improvement by systematically addressing the underlying causes of defects or problems, thereby improving processes and outcomes (Basuki and Fahadha 2020).

2.4.2 Pareto Diagram

The Pareto diagram is an analytical tool for identifying and prioritizing a dataset's most critical problems or causes (Garcia-Bernabeu et al. 2019). The Pareto diagram is based on the principle known as the 80/20 rule, which states that approximately 80% of the effects come from 20% of the causes (Armijal et al. 2023). The Pareto diagram allows this distribution to be visualized graphically and facilitates decision-making, focusing on the most relevant problems (Rodrigues dos Santos et al. 2018). In summary, the Pareto diagram is a tool that helps identify and prioritize the most relevant problems or causes (Vallejo-Castillo et al. 2020).

The Pareto diagram has several advantages, including the following:

(1) It allows one to clearly and graphically visualize a dataset's most critical causes or problems (Pulido-Rojano et al. 2020).
(2) Helps decision-making by focusing on the most relevant problems (Cedeño et al. 2018).
(3) Helps identify and prioritize areas for improvement in a process or system.
(4) Facilitates communication and understanding of data between different teams or departments (Bertocci et al. 2021).
(5) This allows us to identify which causes or problems quickly contribute the most to an effect or outcome (Pacana and Czerwińska 2021).
(6) Helps maximize the efficient use of resources by focusing on the most impactful areas (Jaqin et al. 2020).
(7) Allows setting improvement goals based on the most important causes or problems. (Pacana and Czerwińska 2021).
(8) This facilitates data analysis by summarizing and organizing information in an orderly manner (Bertocci et al. 2021).

Despite these advantages, the Pareto diagram has some disadvantages, such as those mentioned in what follows.

(1) May generate biases if appropriate data are not selected or if relevant information is omitted (Elena and Gerasimova 2019).
(2) Depending on the nature of the data, it may be difficult to categorize causes or problems into mutually exclusive groups (Deidda 2010).
(3) This does not explain the causal relationship between the identified causes and observed effects (Knop and Ziora 2022).
(4) The interpretation and analysis of the Pareto diagram can vary depending on who uses it, leading to different conclusions and solution approaches (Elena and Gerasimova 2019).

From the previous, it follows that focusing on improving parts supply and assembly equipment efficiency would result in the most significant reduction in delays and improvements in the overall efficiency of the production process. Using the Pareto diagram in a continuous improvement process can help identify and prioritize the most important causes or problems, allowing the efficient allocation of resources and strategic focus on critical areas for significant improvements (Bertocci et al. 2021). In summary, the Pareto diagram effectively identifies the most critical problems or causes, prioritizing the actions to be taken (Knop and Ziora 2022).

2.4.3 Statistical Analysis

Statistical analysis is not a technique but rather a set of techniques that allows the analysis of the information obtained. This book does not intend to discuss each technique in-depth, but some of the most important ones and their uses in the DMAIC implementation process are listed in what follows (Chyon et al. 2020):

(1) Normality checks can be performed to determine the type of techniques to be used, either parametric or non-parametric. This technique evaluates whether the data follows a normal distribution, which is crucial for various statistical analyses. Among the most common are the Shapiro-Wilk, Kolmogorov-Smirnov, Anderson-Darling, and Lilliefors tests.
(2) Control charts to monitor process performance over time and detect any variations that may affect quality. There are charts for variables and attributes. Charts for variables can be an X-R chart for the mean (X) and range (R), an X-S chart for standard deviation (S) instead of range, and an individual (I) chart for individual

values of each measurement. The most common attribute p-plots are the p-plot for the proportion of defective units in each sample np-plot, which shows the number of defective units in each sample c-plot, which shows the number of defects per unit of product, and u-plot, which shows the number of defects per unit area. The other graphs are XmR, Shewhart, and CUSUM graphs.

(3) Capability analysis to meet specifications and provide information on process performance.
(4) Classification and regression tree (CART) for predictive modeling and identification of relationships between variables.
(5) Full Factorial Design of Experiments (FFDE) to study the effects of multiple factors on the process output.
(6) MANOVA (multivariate analysis of variance) to analyze differences between group means in multivariate data.

There are many advantages to performing a statistical analysis of the information obtained in DMAIC projects, including the following:

(1) Data-driven decision-making allows organizations to base their improvement initiatives on objective data rather than assumptions (Powell et al. 2017).
(2) Identify root causes that should be addressed immediately, avoiding wasting effort and resources on trivial causes (Jamil et al. 2020a).
(3) Process improvement, by identifying areas for improvement, leads to greater process efficiency and effectiveness (Trubetskaya et al. 2024).
(4) Sustainability by identifying opportunities for waste reduction and resource optimization (Gholami et al. 2021).
(5) Quality can be improved by measuring and monitoring process performance, reducing defects, and increasing customer satisfaction (Ubaid and Dweiri 2024).

However, performing statistical analyses is not easy and is often the least performed in companies because of the level of knowledge required. Some authors, such as Ubaid and Dweiri (2024), mention some disadvantages of statistical analysis, including the following.

(1) Statistical analysis techniques are complex and require specialized knowledge and skills for practical application, which may be difficult for some people.
(2) Statistical analysis depends on the quality of the data obtained from the processes and is often inaccurate or incomplete, which can lead to incorrect conclusions.

(3) Implementing statistical analysis techniques may require significant resources regarding time, training, and specialized software tools, which could be a barrier in small organizations.
(4) Employees unfamiliar with or hesitant to use statistical methods may resist the introduction of statistical analyses and data-driven decision-making approaches.
(5) Cultural barriers within an organization, associated with a lack of understanding of statistical analysis, can obstruct successful implementation.

2.4.4 Analysis of Variance

Analysis of variance (ANOVA) is a statistical technique used to compare the means of two or more groups and determine whether there is any significant difference between them based on the variance of the data (Abdullahi et al. 2019). This analysis is based on the null hypothesis that there is no difference between group means and the alternative hypothesis that at least one mean is different (Eheart et al. 1955). The method described by Snedecor was used to perform the ANOVA. This method involves calculating between-group and within-group variances. The between-group variance is then compared with the within-group variance using a statistical test, generally known as an F-test. Suppose that between-group variance is significantly greater than within-group variance. In this case, the null hypothesis is rejected, and it is concluded that there are significant differences between the means of the groups (Abdullahi et al. 2019).

To understand and interpret the ANOVA results, it is essential to consider factors such as the homogeneity of variance and normality of the data distribution. The assumption of homogeneity of variance can be examined by tests such as Levene's test, which ensures that the variance is consistent between the groups being compared (Şanal 2023). In addition, the normality of the data distribution can be assessed using the Shapiro-Wilk test, which allows us to determine whether the data follow a normal distribution (Vencúrik et al. 2021).

Similarly, post hoc analyses like Tukey's test investigate group differences (Barczyk-Pawelec et al. 2022). This robustness test can identify specific pairs of groups that exhibit significant differences in environmental and ecological parameters. In ANOVA, it is crucial to detect significant differences and elucidate where they reside, thus contributing to an overall understanding of the data (Tian et al. 2020).

ANOVA offers the following advantages:

(1) This allows simultaneous analysis of the differences between several groups or treatments (Nagamine et al. 2009). Thus, the influence of different variables or factors on a variable of interest can be examined more efficiently.

(2) Controlling for confounding variables by including covariates (Faria et al. 2018). This is especially useful when investigating the effect of a treatment or intervention while controlling for other factors that could influence the results.
(3) The relative importance of each variable in the final model can be evaluated (Cao et al. 2019).
(4) It provides measures of dispersion and variability within each group, which helps to understand the variability of the data and detect possible differences between groups (Wang et al. 2022).
(5) It is relatively easy to perform and provides robust statistical results (Gabriel et al. 2016).

However, ANOVA has several disadvantages, such as:

(1) Only complete cases can be analyzed, which can be limited when one wishes to analyze data containing missing values or incomplete data (Stepanek et al. 2022). This can be reduced by imputing missing values from incomplete cases.
(2) Only independent variables were analyzed (Yirga et al. 2020). This can be limited when analyzing the relationship between multiple dependent variables.
(3) ANOVA can be affected by the assumption of homogeneity of variance between the groups, which may not be valid in some cases and may affect the interpretation of the results (Stepanek et al. 2022).

2.5 TOOLS AND TECHNIQUES IN THE IMPROVEMENT PHASE

2.5.1 Brainstorming

Brainstorming is a fundamental component of the DMAIC methodology for implementing process improvements in the industry. This technique generates creative ideas and solutions to problems identified during the DMAIC process, which comes from a working group. Several studies have highlighted the importance of brainstorming in the DMAIC framework. For example, Moszyk and Deja (2023) used brainstorming to identify problems associated with the service times for external trucks at a container terminal. Similarly, Azwir (2022) combined brainstorming with other methods, such as Kaizen and FMEA, to improve the finishing process of food packaging products. In addition, Ricciardi et al. (2019) identified brainstorming as one of the tools used during the DMAIC cycle phases to reduce the length of stay in a clinical setting, as it prioritizes cases.

Combining brainstorming with the DMAIC methodology allows teams to explore diverse perspectives, identify root causes of problems, and develop innovative solutions that are proposed openly and transparently to all working group members. Thus, by fostering open communication and idea generation, brainstorming improves the problem-solving process and fosters a collaborative environment conducive to continuous improvement efforts, consistent with the basic principles of Six Sigma and Lean methodologies, which aim to streamline processes, reduce defects, and optimize performance.

Because brainstorming is conducted, it is essential to highlight several advantages that are often observed when implementing the DMAIC methodology (Mauluddiyah et al. 2018), such as the following:

(1) Brainstorming allows team members with varied backgrounds and experiences to contribute ideas, leading to a thorough exploration of the possible solutions to the problem under analysis.
(2) By fostering a collaborative environment in which all ideas are welcome from all constituents of various departments, brainstorming promotes creative thinking and innovative problem-solving approaches within the holistic framework, as the needs of all parties involved are integrated.
(3) The interactive nature of brainstorming sessions encourages participation and engagement of the entire work team, which facilitates the identification of root causes and the development of practical solutions to continuous process improvement challenges.
(4) Brainstorming promotes team teamwork and collaboration, improving communication, trust, and camaraderie. These are essential for successfully implementing DMAIC, as everyone must be highly committed to problem-solving.
(5) Therefore, through brainstorming, organizations can instill a culture of continuous improvement by encouraging idea generation, problem-solving, and decision-making, aligning with the principles of Six Sigma and Lean methodologies.

However, brainstorming also has several disadvantages that should be considered to avoid them. Jindal and Maini (2022) mention the following:

(1) Brainstorming sessions sometimes lead to groupthink, where specific individuals conform to dominant ideas within the group, potentially limiting creativity and overlooking alternative perspectives. This also occurs when ideas come from people high in the company hierarchy.

(2) Brainstorming sessions can be time-consuming and can be used to produce, primarily if not properly managed, inefficiencies in the DMAIC process and delay decision-making and implementation of solutions, running the risk of getting into debates between individuals or groups of individuals.
(3) Certain team members may sometimes dominate the brainstorming session, whereas others may be less inclined to contribute. This can result in uneven participation and potentially missing valuable ideas from people who genuinely know the problems to be solved.
(4) Without proper facilitation and structure, brainstorming sessions can lack focus and direction, generating ideas that are not directly relevant to the problem and do not focus on continuous improvement.
(5) Brainstorming often focuses on the generation of ideas. However, it may lack robust mechanisms for evaluating and prioritizing the ideas generated, potentially resulting in the selection of suboptimal solutions for process improvement within the DMAIC framework.

2.5.2 Failure Mode and Effect Analysis (FMEA)

FMEA is a systematic methodology in the framework of DMAIC implementation. It allows the identification of possible failure modes in processes, systems, or products and assessing their impact on the overall performance. Therefore, the objective of FMEA is to anticipate and address risks proactively before they occur, thus improving the quality, reliability, and safety of the production processes. Through FMEA, the industry can prioritize strategies to mitigate risks, allocate resources efficiently, and prevent or minimize failures that compromise operational efficiency or product quality.

Owing to its structured approach, the FMEA allows the identification of possible failure modes, assessment of the severity of their effects, and determination of the probability of their occurrence. This facilitates the evaluation of the capacity to detect or mitigate risks and to promote their management. Thus, FMEA provides a framework for understanding vulnerabilities within a company, improving product design, optimizing processes, and ensuring compliance with quality standards and regulatory requirements.

This leads to the conclusion that applying the FMEA approach offers certain advantages, some of which are listed in what follows.

(1) FMEA allows potential failure modes in production processes to be proactively identified and addressed, enabling the implementation of preventive measures to mitigate risks before they directly impact the operations (Tsai et al. 2017).

(2) By systematically analyzing failure modes and their effects, FMEA helps improve design before series production, improving reliability and product performance in the market and leading to products that meet customer expectations (Mi et al. 2018).
(3) FMEA helps identify critical failure modes in the early design stages, thus achieving better process planning. This reduces costs by avoiding waste generation, rework, or corrections, reducing warranty claims, and minimizing downtime due to failures.
(4) The structured FMEA approach provides information that supports informed decision-making, resource allocation, and prioritization of strategies to mitigate risks, leading to more effective and efficient processes.
(5) By improving product reliability, safety, and quality, FMEA promotes and supports improved competitiveness in the marketplace, builds customer confidence, and differentiates itself from the competition.

However, FMEA may also have disadvantages when applied in the DMAIC environment, such as the following:

(1) FMEA often relies on subjective judgments from experts, which can introduce biases and inconsistencies in identifying and assessing failure modes and their effects, leading to a focus on trivial causes (Sharma et al. 2005).
(2) Conducting FMEA requires time, expertise, and scarce resources, which can be a challenge for small companies looking for short-term solutions (Shafiee and Dinmohammadi 2014).
(3) FMEA can focus on known failure modes and overlook rare or unforeseen failure scenarios that are sometimes not understood, potentially leaving critical risks unaddressed and focusing on trivial causes (Peeters et al. 2018).
(4) The complexity of FMEA methodologies requires expertise in several areas of knowledge, such as mechanical design and specialized software. In addition, operations and maintenance can make the implementation process challenging and time-consuming (Arabian-Hoseynabadi et al. 2010).
(5) Implementing FMEA in complex systems is frequently challenging owing to high resource requirements, lack of available data, and difficulties in enterprise-wide deployment (Liu et al. 2014).

2.5.3 Design of Experiments (DOE)

Design of experiments (DOE) is a statistical technique used to plan, conduct, and analyze experiments efficiently, reliably, and controllably. DOE seeks to identify and quantify the causal relationships between the variables being studied in an experiment; therefore, it is necessary to determine which variables affect a response (dependent) variable, measure the strength of the relationship between the variables, and identify the optimal conditions for obtaining a desired outcome.

This technique allows control and manipulation of the independent variables to observe their effect on the response variable and, therefore, requires randomization to avoid bias and facilitate the repetition of events. Some of the most common types are entirely randomized designs, where treatments are assigned completely randomly; block designs, where experimental units are grouped into blocks and treatments are randomly assigned to blocks; and factorial designs, which are used to study the effects of multiple independent variables simultaneously.

Because of the statistical rigor of DOE and the identification of relationships between variables, they have the following advantages:

(1) DOE helps identify critical process parameters and their optimum settings or levels, improving process efficiency and product quality and avoiding trivial variables (Assegehegn et al. 2020).
(2) By systematically planning experiments, DOE allows efficient resource allocation, reduces costs, and minimizes the number of trials required to achieve the desired process optimization results (Arboretti et al. 2021).
(3) DOE's structured approach provides organizations with reliable data and insights to make informed decisions based on rigorous analysis (Souza et al. 2019).
(4) DOE helps identify critical factors that significantly impact process performance, allowing organizations to focus on addressing them to improve quality and customer specifications (Driel et al. 2018).
(5) Integrating DOE into industrial processes supports a culture of continuous improvement by systematically evaluating and optimizing processes based on empirical data and statistical analyses directly from production lines (Gremyr et al. 2003).

However, because of its statistical rigor and the need for production line information, it has the following disadvantages (Pepper and Spedding 2010):

(1) DOE can be complex and time-consuming to set up and run, as it requires thorough knowledge of the statistics.
(2) DOE implementation in DMAIC requires additional specialized personnel, equipment, and time resources, increasing the improvement initiative's cost.
(3) DOE is not always suitable for all types of processes or projects within the DMAIC methodology, limiting its applicability to specific scenarios in which the variables can be controlled.

(4) Analyzing and correctly interpreting the DOE results is challenging, especially for people without a solid statistical background, and can lead to incorrect conclusions and decisions.
(5) Integrating DOE properly into the DMAIC process poses challenges in aligning the experimental design with the specific objectives and requirements of each DMAIC phase, where it is sometimes difficult to identify all variables involved.

2.5.4 Pilot Programs

A pilot program refers to small-scale testing of process improvements before their full-scale implementation on production lines. As such, the primary objective of a pilot program is to evaluate the feasibility, effectiveness, and potential challenges of the proposed changes in a controlled environment that allows the monitoring of the behavior of variables. This allows managers and engineers to collect data, identify problems, and refine improvement strategies before a more comprehensive implementation, which, in turn, allows for adjustments based on empirical evidence to ensure successful implementation and minimize risks (Rifqi et al. 2021).

Therefore, in DMAIC, a pilot program serves as a structured method to validate proposed changes, test their impact on processes, and evaluate the effectiveness of improvement initiatives. It provides a controlled environment to monitor and measure results, allowing organizations to make informed decisions before deploying changes on a larger scale. These practices build trust among stakeholders, ensure employee buy-in, and facilitate the smoother implementation of change (Lancaster et al. 2004).

Pilot-testing applications in the industry have been widely reported. For example, pharmaceutical companies have tested the implementation of continuous-flow technology with pilot testing for mass-scale implementation (Escribà-Gelonch et al. 2019). In addition, in the ethanol supply chain, the impact of scaling up a stillage pilot plant to improve critical processes was evaluated, demonstrating that the successful application of a pilot plant results in improving industrial processes (Ramos-Hernández et al. 2016).

The use of pilot programs offers several advantages in the DMAIC methodology that increase the probability of success of process improvement initiatives, including the following (Panayiotou et al. 2022):

(1) Pilot programs evaluate the feasibility of the proposed process improvements in a controlled environment before full-scale implementation, which helps identify challenges and refine the strategies.
(2) By conducting pilot tests, organizations can mitigate the risks associated with process changes, identify and address problems on a smaller scale, and avoid waste.

(3) Pilot programs provide an opportunity to engage stakeholders, build trust, and ensure employee buy-in by demonstrating the effectiveness of the proposed improvements in a tangible manner.
(4) Pilot programs allow valuable data collection and knowledge of the impact of process changes, facilitating informed decision-making based on empirical evidence before scaling up improvements.
(5) Through pilot testing, organizations optimize processes, refine methodologies, and ensure that process improvements align with organizational objectives.

However, implementing pilot programs is difficult and has several disadvantages, including the following (Júnior et al. 2020):

(1) Pilot programs do not always capture all the complexities and variabilities of the processes because they seek to synthesize them. This can lead to problems when extending improvements to the entire organization owing to the omission of critical variables that were not initially considered in this way.
(2) Pilot testing requires additional resources in terms of time, personnel, and equipment, which increases the overall cost of process improvement initiatives owing to investments in technology and personnel.
(3) Analyzing and interpreting data from pilot programs can be complex when extrapolating results to broader organizational contexts. This can lead to misinterpretation of findings and the risk of the run being unrepresentative.
(4) Integrating the pilot program results into the DMAIC process often presents problems in aligning the results with each methodology phase's specific objectives and requirements, where analyses are required for different variables.
(5) Resistance from stakeholders, including employees and management, to the changes identified during pilot testing may hinder the successful implementation of process improvements on a larger scale.

2.6 TOOLS AND TECHNIQUES FOR THE CONTROL PHASE

2.6.1 Control Charts

When integrated into the DMAIC methodology, a control chart is a tool for monitoring and managing process performance. Its purpose is to provide a visual representation of the process data to detect variations, trends, and patterns in performance. Using statistical process control charts, managers can identify when a process operates within acceptable limits and when deviations occur. This allows timely intervention and corrective actions to maintain process stability and quality (Gupta and Kumar 2014).

Therefore, it is concluded that control charts play a crucial role in Six Sigma and DMAIC projects that focus on process improvements. They help monitor process stability, identify sources of variation, and make data-driven decisions to improve the process performance. Thus, managers can effectively track process metrics, measure the impact of improvements, and ensure sustained quality and efficiency in operations (Cano et al. 2021).

As the name implies, control charts are used in the control stage. They focus on monitoring the improvements implemented to maintain gains and ensure corrective action is taken when necessary (Gupta and Kumar 2014). Control charts provide a systematic approach to continuously monitoring the performance achieved, allowing variations to be proactively managed and control maintained over time.

We do not go into these control charts in depth in this section because many have been discussed in the statistical analysis section developed previously. However, we do go a little deeper into the advantages and benefits offered by their implementation:

(1) Control charts provide a visual representation of process data, allowing it to be monitored in real-time, indicating its performance, and facilitating the detection of variations or trends (Swarnakar and Vinodh 2016).
(2) By analyzing and representing data in control charts, managers make informed decisions based on rigorous and properly validated statistical analysis, leading to more effective problem-solving and process optimization (Graham et al. 2011).
(3) Control charts assist in the early detection of process deviations or anomalies, allowing timely intervention and corrective actions to prevent quality problems that, if undetected, can generate waste and reject production batches (Abid et al. 2017).
(4) They facilitate continuous improvement efforts by systematically tracking process metrics, measuring the impact of changes, and driving sustainable quality improvements (Prashar 2014). Such monitoring is often performed in real-time.
(5) They allow for assessing processes' capability, identifying areas for improvement, and evaluating the effectiveness of implemented solutions, leading to improved operational performance (Nabhani and Shokri 2009).

Applying control charts is difficult, and their implementation is often abandoned for the following reasons.

(1) Control charts are often complex to set up and interpret, as they require a solid understanding of the statistical principles and data analysis techniques. They may be exclusively used by a certain number of people within the company (Jastania et al. 2017).
(2) Additional resources are required for specialized personnel, training, and software for rapid and efficient generation, which may increase the cost of this type of improvement initiative (Cano et al. 2021).
(3) Analyzing control chart data correctly can be challenging, especially for people without a strong statistical background. This can lead to possible misinterpretation of the results and, often, subjective interpretations, leading to discussions that confuse users (Kumar et al. 2021).
(4) Implementing and maintaining control charts can be time-consuming, which can delay the identification and resolution of process problems since sufficient information is required to identify patterns (Condé et al. 2023).
(5) Employees and management may resist implementing control charts when they do not understand them correctly, thereby hindering the adoption of data-driven decision-making processes (Wassan et al. 2022).
(6) However, they may not capture all aspects of process performance, potentially overlooking critical factors that affect process quality and efficiency. Hence, it is essential to correctly identify the variables to be monitored and plotted (Gupta et al. 2018).
(7) Relying solely on control chart data may overlook the qualitative aspects of process improvement, such as employee feedback and organizational culture, where previous experience has been had and the consequences are known (Liu et al. 2023).
(8) Like other tools, seamlessly integrating control charts into existing processes and systems can pose challenges, requiring significant coordination and alignment primarily because of the statistical rigor required (Prashar 2014).
(9) Effective implementation of control charts may require the training of employees in statistical analysis and interpretation, which adds to the complexity of the process and the limited time available for such processes, as personnel must leave the production lines (Legesse and Geremew 2021).
(10) Control charts focus on past performance, which may not always accurately predict future process behavior, leading to potential shortcomings in forecasting and decision-making (Hardy et al. 2021). Errors in the data are always possible, which may be due to the measurement equipment or the system itself.

2.6.2 Standard Operating Procedures (SOPs)

Standard Operating Procedures (SOPs) are documents that describe step-by-step instructions for performing a specific task or process within an organization. In other words, they are guidelines that establish the correct way to execute an activity, ensuring consistency in execution, efficiency, and the quality of outcomes (Byrne et al. 2021).

SOPs must comply with a series of characteristics that facilitate their understanding as they aim to improve the process (Byrne et al. 2021):

(1) They should be written in a clear and concise language easily understood by all users, mainly machine and tool operators.
(2) The information should be accurate and reflect current processes and practices. If there are changes in the models and machines, SOPs should be updated.
(3) They should cover all relevant steps and details of the process without omitting important information that would leave operators in doubt about what to do.
(4) Although they establish a standard way of performing the task, they must also be flexible enough to adapt to changes or exceptional situations and the operators' needs.

The application of SOPs in industry offers several advantages, such as the following:

(1) SOPs establish a uniform approach to DMAIC projects, ensuring that all team members adhere to standardized procedures and methodologies throughout the project's life cycle (Swarnakar and Vinodh 2016).
(2) SOPs provide clear instructions on executing activities within each DMAIC phase, providing guidance on roles, responsibilities, and the sequence of tasks to be performed, which defines it as a properly structured methodology (Prashar 2014).
(3) By following SOPs, organizations and their managers maintain quality standards, reduce errors, and ensure that DMAIC projects meet predefined quality criteria, allowing everyone to understand the same procedure (Kumar et al. 2021).
(4) SOPs streamline project execution by eliminating ambiguity in instructions, minimizing rework, and saving time that would otherwise be spent on decision-making and planning. This reduces training time for new operators (Condé et al. 2023).
(5) Since SOPs are standard operating procedures, they ensure that DMAIC projects conform to regulatory requirements, industry

standards, and organizational policies, making the process auditable and transparent and, consequently, organizational compliance and accountability (Ahmed et al. 2023).

(6) SOPs facilitate risk management, including preventive measures, contingency plans, and protocols, to address unforeseen challenges (Acero et al. 2019).

(7) SOPs enable the training of new team members, allowing rapid onboarding, understanding of the DMAIC methodology, and influential contributions to project success (Ponsiglione et al. 2021).

(8) SOPs improve communication and collaboration among project team members by providing a common point of reference, ensuring alignment, and sharing objectives (Antony et al. 2005). However, they act as a repository of knowledge and best practices, allowing organizations to capture lessons learned, share ideas, and build institutional knowledge for future DMAIC projects (Ciasullo et al. 2024).

However, it is also possible to have disadvantages, among which are mentioned in what follows:

(1) DMAIC SOPs are sometimes rigid and inflexible. They may not allow the adaptability and agility needed to address unique challenges or unexpected situations during the process improvement journey (Powell et al. 2017).

(2) SOPs are sometimes complex for operators, leading to confusion among team members and making it challenging to follow prescribed procedures accurately. This complexity hinders the smooth execution of DMAIC projects and may result in delays or errors due to discussions about understanding (Dora and Gellynck 2015).

(3) Implementation of SOPs may face resistance from employees who are used to working in a specific way, which impedes the success of DMAIC initiatives (Scala et al. 2021).

(4) Introducing SOPs requires training to ensure that all team members understand and adhere to procedures. Such training has the disadvantage of being time- and resource-consuming, which affects the overall efficiency of the DMAIC projects (Ponsiglione et al. 2021).

(5) Moreover, relying too much on standard operating procedures can lead to a focus on subsequent procedures rather than encouraging critical thinking and innovation on the part of the work team. This stifles creativity and limits the exploration of alternative solutions when implementing DMAIC (Tsarouhas and Sidiropoulou 2024).

2.6.3 Poka-Yoke

The word Poka-Yoke comes from Japanese and means "error-proof," so it is defined as a quality management technique. Thus, Poka-Yoke aimed to prevent errors in production processes by incorporating mechanisms or devices that detect and eliminate them before they occur. By proactively addressing potential errors, it aims to improve process efficiency, product quality, and overall operational performance (Vinod et al. 2015).

Several well-known companies have implemented this method. For example, Toyota is famous for the extensive use of Poka-Yoke in its manufacturing processes, as it has integrated error-proof devices and mechanisms to avoid defects and ensure high quality in automobile manufacturing. Amazon implemented Poka-Yoke strategies using automated processes to minimize errors in order picking, packing, and shipping processes.

Similarly, Ford Motor Company has adopted Poka-Yoke techniques to improve quality control and error prevention in vehicle assembly processes, which has enabled it to improve production efficiency and product reliability. Finally, Boeing used the Poka-Yoke method in aircraft assembly and production to ensure the precision and accuracy of complex manufacturing processes and to improve safety and quality standards in manufacturing.

The application of Poka-Yoke in industry and the DMAIC methodology offers several benefits for companies and managers, including the following.

(1) Preventing errors and defects at their source before they occur improves the quality of the production process and reduces rework (Jirasukprasert et al. 2015).
(2) By eliminating errors and failures, Poka-Yoke increases productivity and operational efficiency within processes as resources are better used (Saurin et al. 2012).
(3) Poka-Yoke implementation reduces production costs by minimizing errors, defects, and the associated rework or scrap costs (E.V et al. 2019).
(4) By avoiding waste and rework, Poka-Yoke ensures higher product quality (Trakulsunti et al. 2021).
(5) Poka-Yoke generates higher customer satisfaction by reducing errors and improving product quality (Prashar 2014).
(6) Helps streamline processes as errors decrease bottlenecks and inefficiencies, resulting in smoother production systems (Panayiotou et al. 2022).
(7) By providing tools and systems for error prevention, Poka-Yoke empowers employees to take ownership of quality and process improvement. They are the ones proposing new ideas for process improvements, which often reduces risks in the production system.

However, some authors have also observed several disadvantages of applying Poka-Yoke on production lines, including the following:

(1) Poka-Yoke requires an initial investment to generate devices, systems, or error-proof training, which increases the product's production costs. In addition, machines and equipment must sometimes be stopped for adjustments, representing downtime and idle time (Kumar and McKewan 2011).
(2) Sometimes, implementing Poka-Yoke is difficult and complex because it is difficult to design the devices and bring them to an optimal performance level (Huertas-Reyes et al. 2022).
(3) Poka-Yoke devices require periodic maintenance, calibration, or upgrades to ensure their effectiveness, which adds to the ongoing operational costs of shutting down the machinery or equipment on which it is installed (Prabowo and Aisyah 2020).
(4) Implementing Poka-Yoke represents changing work methods, so employees may resist it significantly if it disrupts established workflows or processes with which they are already familiar, hindering adoption and efficiency (Kumar et al. 2021).
(5) Relying too heavily on technology-based Poka-Yoke solutions can lead to a false sense of security and overlook human factors or process nuances that technology may not adequately address (Prashar 2014).
(6) Poka-Yoke solutions often need more flexibility to adapt to changing process requirements. This often leads to arguments between operators and managers, minimizing fluid and constructive communication.

2.7 CONCLUSIONS

The DMAIC methodology focuses on solving problems in the industry, which may differ from one company to another, so it is impossible to indicate a list of tools or techniques that should be used. This chapter has described the most important tools when using the DMAIC methodology and has observed that they offer many advantages but also many disadvantages.

One disadvantage of many of the tools was that they required certain levels of statistical knowledge, which is sometimes not easy. In addition, many of the tools tend to simplify the problems, so companies must ensure that they integrate all the variables that affect the problem they are solving.

However, there is a consensus that the tools used in DMAIC offer an analysis capability that identifies deficiencies and areas for improvement in the production areas. Also, because it already has a structure, DMAIC is an industrially accepted way to solve problems, and only a series of steps and procedures need to be followed. Some of these tools are easy to apply, but others are not.

For this reason, it is recommended that the best way to define the problems to be solved is to identify the best tools since it is not possible to generate a list of which tools will be used generically. If companies do not have sufficient experience in the application of DMAIC tools, they should seek advice from external entities and begin to train workers in them.

REFERENCES

Abdel-Jaber O, Itani A, Al-Hussein M (2022) Hybrid lean decision-making framework integrating value stream mapping and simulation: A manufacturing case study. In: Proceeding of the 30th Annual Conference of the International Group for Lean Construction (IGLC), Edmonton, 27 July, pp. 153–163. https://doi.org/10.24928/2022/0118

Abdullahi SR, Raşit Öner M, Nuri Ata O (2019) Fractional Factorial Design Application for the Determination of Parameters Affecting KOH and HCl Generation from Simulated Wastewater Solution by Bipolar Membrane Electrodialysis. Eur J Sci Technol 866–873. https://doi.org/10.31590/ejosat.646850

Abid M, Nazir HZ, Riaz M, et al. (2017) An efficient nonparametric EWMA Wilcoxon signed-rank chart for monitoring location. Qual Reliab Eng Int 33(3):669–685. https://doi.org/10.1002/qre.2048

Acero R, Torralba M, Pérez-Moya R, et al. (2019) Order processing improvement in military logistics by value stream analysis lean methodology. In: Barajas C, Caja J, Calvo R, et al. (eds) Procedia Manufacturing. Elsevier B.V., pp. 74–81.

Adeodu AO, Kanakana-Katumba MG, Maladzhi R (2020) Implementation of Lean Six Sigma (LSS) methodology, through DMAIC approach to resolve down time process; a case of a paper manufacturing company. In: Proceedings of the International Conference on Industrial Engineering and Operations Management. IEOM Society, pp. 37–47.

Ahammed R, Hasan MZ (2020) Humming noise reduction of ceiling fan in the mass production applying DMAIC-Six Sigma approach. World J Eng 18(1):106–121. https://doi.org/10.1108/wje-07-2020-0329

Ahmed A, Olsen J, Page J (2023) Integration of Six Sigma and simulations in real production factory to improve performance—a case study analysis. Int J Lean Six Sigma 14(2):451–482. https://doi.org/10.1108/IJLSS-06-2021-0104

Aichouni ABE, Abdullah H, Ramlie F (2021) A scientific approach of using the DMAIC methodology to investigate the effect of cutting tool life on product quality and process economics: A case study of a Saudi manufacturing plant. Eng Technol Appl Sci Res 11(1):6799–6805. https://doi.org/10.48084/etasr.4008

Antony J, Gijo EV, Childe SJ (2012) Case study in Six Sigma methodology: Manufacturing quality improvement and guidance for managers. Prod Plan Control 23(8):624–640. https://doi.org/10.1080/09537287.2011.576404

Antony J, Kumar M, Tiwari MK (2005) An application of Six Sigma methodology to reduce the engine-overheating problem in an automotive company. Proc Inst Mech Eng B J Eng Manuf 219(8):633–646. https://doi.org/10.1243/095440505x32418

Arabian-Hoseynabadi H, Oraee H, Tavner P (2010) Failure modes and effects analysis (FMEA) for wind turbines. Int J Elec Power Energ Syst 32(7):817–824. https://doi.org/10.1016/j.ijepes.2010.01.019

Araman H, Saleh Y (2023) A case study on implementing Lean Six Sigma: DMAIC methodology in aluminum profiles extrusion process. TQM J 35(2):337–365. https://doi.org/10.1108/TQM-05-2021-0154

Arboretti R, Ceccato R, Pegoraro L, et al. (2021) Design of experiments and machine learning for product innovation: A systematic literature review. Qual Reliab Eng Int 38(2):1131–1156. https://doi.org/10.1002/qre.3025

Armijal, Marlina WA, Hadiguna RA (2023) The evaluation of supply chain risk management on smallholder layer farms. IOP Conf Ser Earth Environ Sci 1182:012082. https://doi.org/10.1088/1755-1315/1182/1/012082

Assegehegn G, Fuente EB, Franco JM, et al. (2020) An experimental-based approach to construct the process design space of a freeze-drying process: An effective tool to design an optimum and robust freeze-drying process for pharmaceuticals. J Pharm Sci 109(1):785–786. https://doi.org/10.1016/j.xphs.2019.07.001

Augustínová E, Daubner M (2014) Achievement of value for money in PPP projects by selection of key performance indicators. In: International Multidisciplinary Scientific GeoConference Surveying Geology and Mining Ecology Management, SGEM. International Multidisciplinary Scientific GeoConference, pp. 3–9. https://doi.org/10.5593/sgem2014B53

Azwir HH (2022) Improving the finishing process of food packaging products using DMAIC method. J Rekayasa Sist Ind 11(2):129–144. https://doi.org/10.26593/jrsi.v11i2.5318.129-144

Barczyk-Pawelec K, Rubajczyk K, Stefańska M, et al (2022) Characteristics of Body Posture in the Sagittal Plane in 8–13-Year-Old Male Athletes Practicing Soccer. Symmetry (Basel) 14:210. https://doi.org/10.3390/SYM14020210

Baro M, Piña M, Reyes A (2023) 6 Sigma and DMAIC method: Basic tool teaching and application for beginning practitioners in automotive assembly. J Optim Ind Eng 16(1):89–96. https://doi.org/10.22094/JOIE.2023.1976488.2029

Barosz P, Gołda G, Kampa A (2020) Efficiency Analysis of Manufacturing Line with Industrial Robots and Human Operators. Appl Sci 2020, Vol 10, Page 2862 10:2862. https://doi.org/10.3390/APP10082862

Basuki M, Fahadha RU (2020) Identification of the causes nata De Coco production defects for quality control. Spek Ind 18(2):175. https://doi.org/10.12928/si.v18i2.14393

Bellucci M, Simoni L, Acuti D, et al. (2019) Stakeholder engagement and dialogic accounting. Account Audit Account J 32(5):1467–1499. https://doi.org/10.1108/AAAJ-09-2017-3158

Bertocci F, Grandoni A, Fidanza M, Berni R (2021) A Guideline for Implementing a Robust Optimization of a Complex Multi-Stage Manufacturing Process. Appl Sci 11:1418. https://doi.org/10.3390/APP11041418

Bilen U, Helvacioglu S, Helvacioglu IH (2022) Evaluation of small and medium sized shipyards productivity with key performance indicators. In: SNAME Maritime Convention, SMC 2022. Society of Naval Architects and Marine Engineers. https://doi.org/10.5957/SMC-2022-120

Budihardjo R, Hadipuro W (2022) Green value stream mapping: A tool for increasing green productivity (The Case of PT. NIC). 4(1):19. https://doi.org/10.24167/jmbe.v4i1.4620

Byrne B, McDermott O, Noonan J (2021) Applying Lean Six Sigma methodology to a pharmaceutical manufacturing facility: A case study. Processes 9(3):550.

Cano JA, Gómez RA, Cortés P (2021) ICT validation in logistics processes: Improvement of distribution processes in a goods sector company. Informatics 8(4):75.

Cano S, Botero L, García-Alcaraz JL, et al. (2020) Key aspects of maturity assessment in lean construction. In: 28th Annual Conference of the International Group for Lean Construction 2020, IGLC 2020. The International Group for Lean Construction, pp. 229–240. https://doi.org/10.24928/2020/0063

Cao W, Wang SL, Fernandez C, et al (2019) A novel adaptive state of charge estimation method of full life cycling lithium-ion batteries based on the multiple parameter optimization. Energy Sci Eng 7:1544–1556. https://doi.org/10.1002/ESE3.362

Cazacu CC, Chiscop F, Cazacu DA (2021) Using IoT to Dynamically Test Smart Connected Devices. In: Cofaru NF, Inţă M (eds) MATEC Web of Conferences 343. 10th International Conference on Manufacturing Science and Education – MSE 2021. EDP Sciences, Sibiu, Romania, pp. 1–9.

Cedeño EAL, Carguachi-Caizatoa JB, Rocha-Hoyos JC (2018) Energy and exergy evaluation in a 1.6L Otto cycle internal combustion engine. Enfoque UTE 9:221–232. https://doi.org/10.29019/ENFOQUEUTE.V9N4.365

Chyon B, Fuad Ahmed, Ahmmed B, et al. (2020) Measuring process capability in a hospital by using Lean Six Sigma tools—a case study in Bangladesh. Global Adv Health Med 9. https://doi.org/10.1177/2164956120962441

Ciasullo MV, Douglas A, Romeo E, et al. (2024) Lean Six Sigma and quality performance in Italian public and private hospitals: A gender perspective. Int J Qual Reliab Manag 41(3):964–989. https://doi.org/10.1108/IJQRM-03-2023-0099

Clancy R, O'Sullivan D, Bruton K (2023) Data-driven quality improvement approach to reducing waste in manufacturing. TQM J 35(1):51–72. https://doi.org/10.1108/TQM-02-2021-0061

Dasandara M, Dissanayake P, Fernando DJ (2022) Key performance indicators for measuring performance of facilities management services in hotel buildings: A study from Sri Lanka. Facilities 40(5–6):316–332. https://doi.org/10.1108/F-02-2021-0009

Daudelin DH, Ruthazer R, Kwong M, et al. (2020) Stakeholder engagement in methodological research: Development of a clinical decision support tool. J Clin Transl Sci 4(2):133–140. https://doi.org/10.1017/cts.2019.443

De Steur H, Wesana J, Dora MK, et al. (2016) Applying value stream mapping to reduce food losses and wastes in supply chains: A systematic review. Waste Manag 58:359–368. https://doi.org/10.1016/j.wasman.2016.08.025

Deidda R (2010) A multiple threshold method for fitting the generalized Pareto distribution to rainfall time series. Hydrol Earth Syst Sci 14:2559–2575. https://doi.org/10.5194/HESS-14-2559-2010

Docekalová MP, Kocmanová A, Simberová I, et al. (2018) Modelling of social key performance indicators of corporate sustainability performance. Acta Univ Agric Silvic Mendelianae Brun 66(1):303–312. https://doi.org/10.11118/actaun201866010303

Dora M, Gellynck X (2015) Lean Six Sigma implementation in a food processing SME: A case study. Qual Reliab Eng Int 31(7):1151–1159. https://doi.org/10.1002/qre.1852

Driel BA, Berg KJ, Smout M, et al. (2018) Investigating the effect of artists' paint formulation on degradation rates of TiO_2-based oil paints. Herit Sci 6(1). https://doi.org/10.1186/s40494-018-0185-2

Duc ML, Bilik P, Truong TD (2022) Design of Industrial System Using Digital Numerical Control. Qual Innov Prosper 26:135–150. https://doi.org/10.12776/QIP.V26I3.1747

Eheart JP, Young RW, Massey PH, Havis JR (1955) Crop, Light Intensity, Soil pH, and Minor Element Effects on the Yield and Vitamin Content of Turnip Greens. J Food Sci 20:575–581. https://doi.org/10.1111/j.1365-2621.1955.tb16871.x

Elena B, Gerasimova IG (2019) Land Plot Selection Rationale for the Location of Linear Facilities. Land 8:67. https://doi.org/10.3390/LAND8040067

Escribà-Gelonch M, de Leon Izeppi GA, Kirschneck D, et al. (2019) Multistep solvent-free 3 m2 footprint pilot miniplant for the synthesis of annual half-ton rufinamide precursor. ACS Sustain Chem Eng 7(20):17237–17251. https://doi.org/10.1021/acssuschemeng.9b03931

Faria M, Fuertes I, Prats E, et al (2018) Analysis of the neurotoxic effects of neuropathic organophosphorus compounds in adult zebrafish. Sci Rep 8:1–14. https://doi.org/10.1038/s41598-018-22977-4

Farrukh A, Mathrani S, Taskin N (2020) Investigating the Theoretical Constructs of a Green Lean Six Sigma Approach towards Environmental Sustainability: A Systematic Literature Review and Future Directions. Sustainability 12:8247. https://doi.org/10.3390/SU12198247

Flynn BB, Sakakibara S, Schroeder RG, et al. (1990) Empirical research methods in operations management. J Oper Manag 9(2):250–284. https://doi.org/10.1016/0272-6963(90)90098-X

Gabriel B, Bodenmann G, Beach SRH (2016) Gender Differences in Observed and Perceived Stress and Coping in Couples with a Depressed Partner. Open J Depress 5:7–20. https://doi.org/10.4236/OJD.2016.52002

Gackowiec P, Podobinska-Staniec M, Brzychczy E, et al. (2020) Review of key performance indicators for process monitoring in the mining industry. Energies 13(19). https://doi.org/10.3390/en13195169

Gagné V, Berthelot S, Coulmont M (2022) Stakeholder engagement practices and impression management. J Glob Responsib 13(2):217–241. https://doi.org/10.1108/JGR-03-2021-0036

Garcia-Bernabeu A, Salcedo J V., Hilario A, et al (2019) Computing the mean-variance-sustainability nondominated surface by ev-moga. Complexity 2019:. https://doi.org/10.1155/2019/6095712

Gartlehner G, Schultes M-T, Titscher V, et al. (2017) User testing of an adaptation of fishbone diagrams to depict results of systematic reviews. BMC Med Res Methodol 17(1):169. https://doi.org/10.1186/s12874-017-0452-z

Gholami H, Jamil N, Mat Saman MZ, et al. (2021) The application of Green Lean Six Sigma. Bus Strateg Environ 30(4):1913–1931. https://doi.org/10.1002/bse.2724

Gijo EV, Antony J, Sunder MV (2019) Application of Lean Six Sigma in IT support services—a case study. TQM J 31(3):417–435. https://doi.org/10.1108/TQM-11-2018-0168

Gijo EV, Bhat S, Jnanesh NA (2014) Application of Six Sigma methodology in a small-scale foundry industry. Int J Lean Six Sigma 5(2):193–211. https://doi.org/10.1108/IJLSS-09-2013-0052

González-González H, Escobar-Prado CA (2021) Aplicación de la herramienta SIPOC a la cadena de suministro interna de una empresa distribuidora de medicamentos. Lumen Gentium 5:119–134. https://doi.org/doi.org/10.52525/lg.v5n2a8

Graham MA, Chakraborti S, Human SW (2011) A nonparametric exponentially weighted moving average signed-rank chart for monitoring location. Comput Stat Data Anal 55(8):2490–2503. https://doi.org/10.1016/j.csda.2011.02.013

Gremyr I, Arvidsson M, Johansson P (2003) Robust design methodology: Status in the Swedish Manufacturing Industry. Qual Reliab Eng Int 19(4):285–293. https://doi.org/10.1002/qre.584

Gupta K, Kumar G (2014) Six Sigma application in warehouse for damaged bags: A case study. In: Proceedings of 3rd International Conference on Reliability, Infocom Technologies and Optimization, 8–10 October, pp. 1–6. https://doi.org/10.1109/ICRITO.2014.7014736

Gupta V, Jain R, Meena ML, et al. (2018) Six-Sigma application in tire-manufacturing company: A case study. J Ind Eng Int 14(3):511–520. https://doi.org/10.1007/s40092-017-0234-6

Hamdan A, Hamdan S, Alsyouf I, et al. (2024) Enhancing sustainability performance of universities: A DMAIC approach. Syst Res Behav Sci 41(1):153–172. https://doi.org/10.1002/sres.2942

Hamid A, Baba I, Sani W (2017) Risk management framework in oil field development project by enclosing fishbone analysis. Int J Adv Sci Eng Inf Technol 7(2):446. https://doi.org/10.18517/ijaseit.7.2.1499

Hatmoko JUD, Neilkelvin J (2023) Key performance indicator for analytical hierarchy process used for determining the effect of reverse supply chain toward green building projects. In: Kristiawan SA, Gan BS, Shahin M, et al. (eds) Lecture Notes in Civil Engineering. Springer Science and Business Media Deutschland GmbH, pp. 823–831. https://doi.org/10.1007/978-981-16-9348-9_73

Hekmatpanah M (2011) The application of cause and effect diagram in the oil industry in Iran: The case of four liter oil canning process of sepahan oil company. Afr J Bus Manag 5(26). https://doi.org/10.5897/ajbm11.1517

Hernández-Soto R, Gutiérrez-Ortega M, Rubia-Avi B (2021a) Exploratory Systematic Review on the Exchange of Knowledge in Virtual Communities of Practice. New Trends Qual Res 9:239–248. https://doi.org/10.36367/ntqr.9.2021.239-248

Huertas-Reyes A, Quispe-Huerta S, Leon-Chavarri C, et al. (2022) Increased efficiency of a metalworking SME through process redesign using SMED, Poka Yoke and Work Standardization. In: Larrondo Petrie MM (ed.) 2nd LACCEI International Multiconference on Entrepreneurship, Innovation and Regional Development—LEIRD 2022, Bogota, DC, 2–6 December, pp. 1–6. www.laccei.org; http://dx.doi.org/10.18687/LEIRD2022.1.1.69

Ilori AE, Sawa BA, Gobir AA (2020) Application of cause-and-effect-analysis for evaluating causes of fire disasters in public and private secondary schools in Ilorin Metropolis, Nigeria. Arch Curr Res Int 19(2):1–11. https://doi.org/10.9734/acri/2019/v19i230156

Indra S, Tumanggor OSP, Hardi Purba H (2021) Value stream mapping: Literature review and implications for service industry. J Sist Teknik Ind 23(2):155–166. https://doi.org/10.32734/jsti.v23i2.6038

Jamil N, Gholami H, Saman MZM, et al. (2020) DMAIC-based approach to sustainable value stream mapping: Towards a sustainable manufacturing system. Econ Res Ekon Istraz 33(1):331–360. https://doi.org/10.1080/1331677X.2020.1715236

Jaqin C, Rozak A, Purba HH (2020) Case Study in Increasing Overall Equipment Effectiveness on Progressive Press Machine Using Plan-do-check-act Cycle. Int J Eng 33:2245–2251. https://doi.org/10.5829/IJE.2020.33.11B.16

Jastania RA, Balata GF, Abd El-Hady MIS, et al. (2017) A qualitative study to improve the student learning experience. Qual Assur Educ 25(4):462–474. https://doi.org/10.1108/QAE-06-2016-0031

Jasti NVK, Kota S, Sangwan KS (2020) An application of value stream mapping in auto-ancillary industry: A case study. TQM J 32(1):162–182. https://doi.org/10.1108/TQM-11-2018-0165

Jindal A, Maini N (2022) Six Sigma in blood transfusion services: A dream too big in a third world country? Vox Sang 117(11):1271–1278. https://doi.org/10.1111/vox.13349

Jirasukprasert P, Garza-Reyes JA, Kumar V, et al. (2015) A Six Sigma and DMAIC application for the reduction of defects in a rubber gloves manufacturing process. Int J Lean Six Sigma 5(1):2–22. https://doi.org/10.1108/IJLSS-03-2013-0020

Johansson PEC, Malmsköld L, Fast-Berglund Å, et al. (2020) Challenges of handling assembly information in global manufacturing companies. J Manuf Technol Manag 31(5):955–976. https://doi.org/10.1108/JMTM-05-2018-0137

Júnior WA, Pianno RFC, Achcar JA (2020) Control of the variability of the biofuel packaging process through the Six Sigma methodology: A case study. Indep J Manag Prod 11(6):2009–2031. https://doi.org/10.14807/ijmp.v11i6.1153

Kartinawati A, Risyahadi ST, Bashar FM (2022) Waste reducing efforts in the kitchen area of the hotel industry using lean management (a case study of XYZ hotel in Bogor). In: Farobie O, Soma T, Arkeman Y, et al. (eds) E3S Web of Conferences. EDP Sciences, Bogor, Indonesia, pp. 1–14.

Kaur A, Lodhia S (2018) Stakeholder engagement in sustainability accounting and reporting. Account Audit Account J 31(1):338–368. https://doi.org/10.1108/AAAJ-12-2014-1901

Knezek EB, Vu T, Lee J (2022) Utilizing DMAIC method to optimize law enforcement official willingness to respond to disasters: An exploratory study. Int J Emerg Serv 11(1):84–113. https://doi.org/10.1108/IJES-11-2020-0068

Knop K, Ziora R (2022) Statistical Analysis and Prediction of the Product Complaints. Syst Saf Hum - Tech Facil - Environ 4:99–115. https://doi.org/10.2478/CZOTO-2022-00011

Krasodomska J, Zarzycka E (2021) Key performance indicators disclosure in the context of the EU directive: When does stakeholder pressure matter? Meditari Account Res 29(7):1–30. https://doi.org/10.1108/MEDAR-05-2020-0876

Kumar S, McKewan GW (2011) Six Sigma DMAIC quality study: Expanded nurse practitioner's role in health care during and posthospitalization within the United States. Home Heal Care Manag Pract 23(4):271–282. https://doi.org/10.1177/1084822310388385

Lancaster GA, Dodd S, Williamson PR (2004) Design and analysis of pilot studies: Recommendations for good practice. J Eval Clin Pract 10(2):307–312. https://doi.org/10.1111/j.2002.384.doc.x

Lastochkina VV (2021) Management on the basis of key performance indicators at manufacturing enterprises. Russ Eng Res 41(12):1193–1195. https://doi.org/10.3103/S1068798X21120273

Laureani A, Antony J (2012) Critical success factors for the effective implementation of Lean Sigma. Int J Lean Six Sigma 3(4):274–283. https://doi.org/10.1108/20401461211284743

Legesse B, Geremew S (2021) Minimizing costs of poor quality for glass container bottles production using Six Sigma's DMAIC methodology: A case study in a bottle and glass share company. Ind Eng 5(1):7–20. https://doi.org/10.11648/j.ie.20210501.12

Li H, Hong T, Lee SH, et al. (2020) System-level key performance indicators for building performance evaluation. Energ Build 209:109703. https://doi.org/10.1016/j.enbuild.2019.109703

Li Y, Wang C, Gao L, et al (2021) An improved simulated annealing algorithm based on residual network for permutation flow shop scheduling. Complex Intell Syst 7:1173–1183. https://doi.org/10.1007/S40747-020-00205-9/TABLES/5

Liliana L (2016) A new model of Ishikawa diagram for quality assessment. IOP Conf Ser Mater Sci Eng 161(1):012099. https://doi.org/10.1088/1757-899X/161/1/012099

Liu HC, Li P, You J, et al. (2014) A novel approach for FMEA: Combination of interval 2-tuple linguistic variables and gray relational analysis. Qual Reliab Eng Int 31(5):761–772. https://doi.org/10.1002/qre.1633

Liu Y, Ren Y, Zhang M, et al. (2023) Solenoid valves quality improvement based on Six Sigma management. Int J Lean Six Sigma 14(1):72–93. https://doi.org/10.1108/IJLSS-08-2021-0140

Lorenzon dos Santos D, Giglio R, Helleno AL, et al. (2019) Environmental aspects in VSM: A study about barriers and drivers. Prod Plan Control 30(15):1239–1249. https://doi.org/10.1080/09537287.2019.1605627

Marques PADA, Matthé R (2017) Six Sigma DMAIC project to improve the performance of an aluminum die casting operation in Portugal. Int J Qual Reliab Manag 34(2):307–330. https://doi.org/10.1108/IJQRM-05-2015-0086

Martínez J, Piersol CV, Holloway S, et al. (2021) Evaluating stakeholder engagement: Stakeholder-centric instrumentation process (SCIP). West J Nurs Res 43(10):949–961. https://doi.org/10.1177/01939459211004274

Maté A, Trujillo J, Mylopoulos J (2017) Specification and derivation of key performance indicators for business analytics: A semantic approach. Data Knowl Eng 108:30–49. https://doi.org/10.1016/j.datak.2016.12.004

Mauluddiyah A, Annisa DI, Sari DFW, et al. (2018) Idea generation on social media based brainstorming session. Atlantis Press, pp. 189–193. https://doi.org/10.2991/miseic-18.2018.46

Memon IA, Ali A, Memon MA, et al. (2019) Controlling the defects of paint shop using seven quality control tools in an automotive factory. Eng Technol Appl Sci Res 9(6):5062–5065. https://doi.org/10.48084/etasr.3160

Mi C, Chen Y, Zhou Z, et al. (2018) Product redesign evaluation: An improved quality function deployment model based on failure modes and effects analysis and 2-tuple linguistic. Adv Mech Eng 10(11). https://doi.org/10.1177/1687814018811227

Moktadir MA, Dwivedi A, Rahman A, et al. (2020) An investigation of key performance indicators for operational excellence towards sustainability in the leather products industry. Bus Strateg Environ 29(8):3331–3351. https://doi.org/10.1002/bse.2575

Morlock F, Meier H (2015) Service value stream mapping in industrial product-service system performance management. Procedia CIRP 30:457–461. https://doi.org/10.1016/j.procir.2015.02.128

Moszyk K, Deja M (2023) Reduction of exceeding the guaranteed service time for external trucks at the DCT Gdańsk container terminal using a Six Sigma framework. Int J Lean Six Sigma 14(7):1566–1595. https://doi.org/10.1108/IJLSS-05-2022-0100

Muhammad U, Ferrer BR, Mohammed WM, et al. (2018) An approach for implementing key performance indicators of a discrete manufacturing simulator based on the ISO 22400 standard. In: Proceedings—2018 IEEE Industrial Cyber-Physical Systems, ICPS 2018. Institute of Electrical and Electronics Engineers, pp. 629–636. https://doi.org/10.1109/ICPHYS.2018.8390779

Mutingi M, Nangolo V, Musiyarira H, et al. (2016) Adoption of maintenance key performance indicators in the Namibian mining industry. In: Grundfest WS, Douglas C, Ao SI (eds) Lecture Notes in Engineering and Computer Science. Newswood Limited, pp. 991–995.

Nabhani F, Shokri A (2009) Reducing the delivery lead time in a food distribution SME through the implementation of Six Sigma methodology. J Manuf Technol Manag 20(7):957–974. https://doi.org/10.1108/17410380910984221

Nagamine Y, Pong-Wong R, Visscher PM, Haley CS (2009) Detection of multiple quantitative trait loci and their pleiotropic effects in outbred pig populations. Genet Sel Evol 41:1–11. https://doi.org/10.1186/1297-9686-41-44/TABLES/8

Noori B, Latifi M (2018) Development of Six Sigma methodology to improve grinding processes. Int J Lean Six Sigma 9(1):50–63. https://doi.org/10.1108/IJLSS-11-2016-0074

Nourbakhsh M, Mydin SH, Zin RM, et al. (2012) Relative importance of key performance indicators of construction projects towards buildability at design stage. Adv Mater Res 340–344. https://doi.org/10.4028/www.scientific.net/AMR.446-449.340

NSB A, Kumar V, De T, Kalangrit S (2022) Quality function deployment analysis of smartphones. Teknomekanik 5:72–79. https://doi.org/10.24036/TEKNOMEKANIK.V5I2.14372

Omair M, Sarkar B, Cárdenas-Barrón LE (2017) Minimum Quantity Lubrication and Carbon Footprint: A Step towards Sustainability. Sustainability 9:714. https://doi.org/10.3390/SU9050714

Pacana A, Czerwińska K (2021) Model of Diagnosing and Searching for Incompatibilities in Aluminium Castings. Materials (Basel) 14:6497. https://doi.org/10.3390/MA14216497

Panayiotou NA, Stergiou KE, Panagiotou N (2022) Using Lean Six Sigma in small and medium-sized enterprises for low-cost/high-effect improvement initiatives: A case study. Int J Qual Reliab Manag 39(5):1104–1132. https://doi.org/10.1108/IJQRM-01-2021-0011

Patyal VS, Koilakuntla M (2016) Relationship between organisational culture, quality practices and performance: Conceptual framework. Int J Product Qual Manag 19(3):319–344. https://doi.org/10.1504/IJPQM.2016.079779

Patyal VS, Modgil S, Koilakuntla M (2021) Application of Six Sigma methodology in an Indian chemical company. Int J Product Perform Manag 70(2):350–375. https://doi.org/10.1108/IJPPM-03-2019-0128

Peeters JJ, Basten R, Tinga T (2018) Improving failure analysis efficiency by combining FTA and FMEA in a recursive manner. Reliab Eng Syst Saf 172:36–44. https://doi.org/10.1016/j.ress.2017.11.024

Pepper MPJ, Spedding TA (2010) The evolution of Lean Six Sigma. Int J Qual Reliab Manag 27 (2):138–155. https://doi.org/10.1108/02656711011014276

Petkova P, Petkova M, Jekov B, et al. (2023) Literature review of key performance indicators for supplier quality management in automotive electronics industry. In: Solic P, Nizetic S, Rodrigues JJPC, et al. (eds) 2023 8th International Conference on Smart and Sustainable Technologies, SpliTech 2023. Institute of Electrical and Electronics Engineers. https://doi.org/10.23919/SpliTech58164.2023.10192584

Ponsiglione AM, Ricciardi C, Scala A, et al. (2021) Application of DMAIC Cycle and Modeling as Tools for Health Technology Assessment in a University Hospital. J Healthc Eng 2021:8826048. https://doi.org/10.1155/2021/8826048

Powell D, Lundeby S, Chabada L, et al. (2017) Lean Six Sigma and environmental sustainability: The case of a Norwegian dairy producer. Int J Lean Six Sigma 8 (1):53–64. https://doi.org/10.1108/IJLSS-06-2015-0024

Prabowo RF, Aisyah S (2020) Poka-Yoke method implementation in industries: A systematic literature review. Indon J Ind Eng Manag 1 (1):12–24

Prashar A (2014) Adoption of Six Sigma DMAIC to reduce cost of poor quality. Int J Product Perform Manag 63(1):103–126. https://doi.org/10.1108/IJPPM-01-2013-0018

Prashar A (2020) Adopting Six Sigma DMAIC for environmental considerations in process industry environment. TQM J 32(6):1241–1261. https://doi.org/10.1108/TQM-09-2019-0226

Pulido-Rojano AD, Ruiz-Lázaro A, Ortiz-Ospino LE (2020) Mejora de procesos de producción a través de la gestión de riesgos y herramientas estadísticas. Ingeniare Rev Chil Ing 28:56–67. https://doi.org/10.4067/S0718-33052020000100056

Ramos-Hernández R, Mota-López DR, Sánchez-Ramírez C, et al. (2016) Assessing the impact of a vinasse pilot plant scale-up on the key processes of the ethanol supply chain. Math Prob Eng 2016:3504682. https://doi.org/10.1155/2016/3504682

Reilly JB, Myers JS, Salvador D, et al. (2014) Use of a novel, modified fishbone diagram to analyze diagnostic errors. Diagnosis 1(2):167–171. https://doi.org/10.1515/dx-2013-0040

Ricciardi C, Fiorillo A, Valente AS, et al. (2019) Lean Six Sigma approach to reduce LOS through a diagnostic-therapeutic-assistance path at A.O.R.N. A. Cardarelli. TQM J 31(5):657–672. https://doi.org/10.1108/TQM-02-2019-0065

Rifqi H, Zamma A, Ben Souda S, et al. (2021) Lean manufacturing implementation through DMAIC approach: A case study in the automotive industry. Qual Innov Prosper 25(2):54–77. https://doi.org/10.12776/qip.v25i2.1576

Rimantho D, Sari IL (2023) Lean manufacturing implementation strategy in the pharmaceutical industry production processes: A VSM and AHP approach. In: Septiani W, Wahyukaton W, Maulidya R, et al. (eds) AIP Conference Proceedings. American Institute of Physics Inc. https://doi.org/10.1063/5.0104932

Rodrigues D, Godina R, da Cruz PE (2021) Key performance indicators selection through an analytic network process model for tooling and die industry. Sustainability 13(24):13777.

Rodrigues dos Santos A, Filho F de SP, Moraes de Almeida F, et al (2018) Innovating Management Control by Dynamic Analysis of Pareto in a Hotel Business. Int J Adv Eng Res Sci 5:197–206. https://doi.org/10.22161/IJAERS.5.4.29

Salwin M, Jacyna-Gołda I, Bańka M, et al. (2021) Using value stream mapping to eliminate waste: A case study of a steel pipe manufacturer. Energies 14(12):3527.

Şanal SÖ (2023) Is more comfortable reading possible with collaborative digital games? An experimental study. J Educ Technol Online Learn 6:116–131. https://doi.org/10.31681/JETOL.1153660

Santosa I, Mulyana R (2023) The IT Services Management Architecture Design for Large and Medium-sized Companies based on ITIL 4 and TOGAF Framework. JOIV Int J Informatics Vis 7:30–36. https://doi.org/10.30630/JOIV.7.1.1590

Sarwar J, Khan AA, Khan A, et al. (2022) Impact of stakeholders on Lean Six Sigma project costs and outcomes during implementation in an air-conditioner manufacturing industry. Processes 10(12). https://doi.org/10.3390/pr10122591

Saurin TA, Ribeiro JLD, Vidor G (2012) A framework for assessing poka-yoke devices. J Manuf Syst 31(3):358–366. https://doi.org/10.1016/j.jmsy.2012.04.001

Scala A, Ponsiglione AM, Loperto I, et al. (2021) Lean Six Sigma approach for reducing length of hospital stay for patients with femur fracture in a University Hospital. Int J Environ Res Public Health 18(6):2843.

Shafiee M, Dinmohammadi F (2014) An FMEA-based risk assessment approach for wind turbine systems: A comparative study of onshore and offshore. Energies 7(2):619–642. https://doi.org/10.3390/en7020619

Shamsu Anuar MA, Mansor MA (2022) Application of value stream mapping in the automotive industry: A case study. J Mod Manuf Syst Technol 6(2):34–41. https://doi.org/10.15282/jmmst.v6i2.8561

Sharma RK, Kumar D, Kumar P (2005) Systematic failure mode effect analysis (FMEA) using fuzzy linguistic modelling. Int J Qual Reliab Manag 22(9):986–1004. https://doi.org/10.1108/02656710510625248

Silvani T, Yanuar A, Juliani W (2019) Application of Six Sigma Method with DMAI Approach in Railway Manufacturing Company. In: Oktavianty O, Eunike A, Lukodono RP, Sari SIK (eds) Proceedings of the 2019 1st International Conference on Engineering and Management in Industrial System (ICOEMIS 2019). Atlantis Press, Malang, Indonesia, pp. 197–204.

Soltani M, Aouag H, Mouss MD (2020) An integrated framework using VSM, AHP and TOPSIS for simplifying the sustainability improvement process in a complex manufacturing process. J Eng Des Technol 18(1):211–229. https://doi.org/10.1108/JEDT-09-2018-0166

Souza LFDS, Vandepitte D, Tita V, et al. (2019) Dynamic response of laminated composites using design of experiments: An experimental and numerical study. Mech Syst Signal Process 115:82–101. https://doi.org/10.1016/j.ymssp.2018.05.022

Stepanek L, Habarta F, Mala I, Marek L (2022) A short note on post-hoc testing using random forests algorithm: Principles, asymptotic time complexity analysis, and beyond. In: Ganzha M, Maciaszek L, Paprzycki M, Ślęzak D (eds) Proceedings of the 17th Conference on Computer Science and Intelligence Systems. FedCSIS, Sofia, Bulgaria, pp. 489–497.

Subagyo IE, Saraswati D, Trilaksono T, et al. (2020) Benefits and challenges of DMAIC methodology implementation in service companies: An exploratory study. J Aplikasi Manajemen 18(4):814–824. https://doi.org/10.21776/ub.jam.2020.018.04.19

Sucipto S, Ariani I, Wulandari S (2022) Continuous quality improvement by statistical process control implementation in cocoa agroindustry. IOP Conf Ser Earth Environ Sci 1024(1):012073. https://doi.org/10.1088/1755-1315/1024/1/012073

Sulthan NS, Jaswadi J, Sulistiono S (2021) Designing Debt Payment Standard Operating Procedures in the SMEs Retail Industry Using Business Process Modeling and Notation (BPMN): A Case Study of a Retailer in East Java, Indonesia. Dinasti Int J Econ Financ Account 2:378–386. https://doi.org/10.38035/DIJEFA.V2I4.992

Sun C, Kang H, Ma H, et al. (2023) A key performance indicator-relevant approach based on kernel entropy component regression model for industrial system. Optim Contr Appl Met 44(3):1540–1555. https://doi.org/10.1002/oca.2770

Sunder MV (2016) Lean Six Sigma project management—a stakeholder management perspective. TQM J 28(1):132–150. https://doi.org/10.1108/TQM-09-2014-0070

Swarnakar V, Vinodh S (2016) Deploying Lean Six Sigma framework in an automotive component manufacturing organization. Int J Lean Six Sigma 7(3):267–293. https://doi.org/10.1108/IJLSS-06-2015-0023

Tambare P, Meshram C, Lee C-C, et al. (2022) Performance measurement system and quality management in data-driven Industry 4.0: A review. Sensors 22(1):224.

Tian J, Han Z, Bogena HR, et al (2020) Estimation of subsurface soil moisture from surface soil moisture in cold mountainous areas. Hydrol Earth Syst Sci 24:4659–4674. https://doi.org/10.5194/HESS-24-4659-2020

Trakulsunti Y, Antony J (2018) Can Lean Six Sigma be used to reduce medication errors in the health-care sector? Leadersh Health Serv 31(4):426–433. https://doi.org/10.1108/LHS-09-2017-0055

Trakulsunti Y, Antony J, Douglas JA (2021) Lean Six Sigma implementation and sustainability roadmap for reducing medication errors in hospitals. TQM J 33(1):33–55. https://doi.org/10.1108/TQM-03-2020-0063

Trubetskaya A, McDermott O, Brophy P (2023) Implementing a customised Lean Six Sigma methodology at a compound animal feed manufacturer in Ireland. Int J Lean Six Sigma 14 (5):1075–1095. https://doi.org/10.1108/IJLSS-08-2022-0169

Trubetskaya A, Ryan A, Powell DJ, et al. (2024) Utilising a hybrid DMAIC/TAM model to optimise annual maintenance shutdown performance in the dairy industry: A case study. Int J Lean Six Sigma 15(8):70–92. https://doi.org/10.1108/IJLSS-05-2023-0083

Tsai SB, Zhou J, Gao Y, et al. (2017) Combining FMEA With DEMATEL models to solve production process problems. PLoS One 12(8):e0183634. https://doi.org/10.1371/journal.pone.0183634

Tsarouhas P, Sidiropoulou N (2024) Application of Six Sigma methodology using DMAIC approach for a packaging olives production system: A case study. Int J Lean Six Sigma 15(2):247–273. https://doi.org/10.1108/IJLSS-06-2022-0140

Tufail MMB, Shamim A, Ali A, et al. (2022) DMAIC methodology for achieving public satisfaction with health departments in various districts of Punjab and optimizing CT scan patient load in urban city hospitals. AIMS Pub Health 9(2):440–457. https://doi.org/10.3934/publichealth.2022030

Ubaid AM, Dweiri FT (2024) Developing an enhanced business process improvement methodology (EBPIM). Int J Lean Six Sigma 15(2):439–468. https://doi.org/10.1108/IJLSS-07-2022-0154

Valier ARS (2020) Looking to Improve Your Practice? Consider the Science of Quality Improvement to Get Started. J Athl Train 55:1137–1141. https://doi.org/10.4085/1062-6050-0342.19

Vallejo-Castillo V, Rodríguez-Stouvenel A, Martínez R, Bernal C (2020) Development of alginate-pectin microcapsules by the extrusion for encapsulation and controlled release of polyphenols from papaya (Carica papaya L.). J Food Biochem 44:e13331. https://doi.org/10.1111/JFBC.13331

Vencúrik T, Knjaz D, Rupčić T, et al (2021) Kinematic Analysis of 2-Point and 3-Point Jump Shot of Elite Young Male and Female Basketball Players. Int J Environ Res Public Heal 18:934. https://doi.org/10.3390/IJERPH18030934

Vendrame Takao MR, Woldt J, da Silva IB (2017) Six Sigma methodology advantages for small- and medium-sized enterprises: A case study in the plumbing industry in the United States. Adv Mech Eng 9(10). https://doi.org/10.1177/1687814017733248

Vinod M, Devadasan SR, Sunil DT, et al. (2015) Six Sigma through Poka-Yoke: A navigation through literature arena. Int J Adv Manuf Technol 81(1):315–327. https://doi.org/10.1007/s00170-015-7217-9

Wang X, Xu Z, Jiang L, et al (2022) Importance Analysis of Structural Seismic Demand Based on Support Vector Machine. Math Probl Eng 2022:1–14. https://doi.org/10.1155/2022/1283105

Wassan RK, Hulio ZH, Gopang MA, et al. (2022) Practical application of Six Sigma methodology to reduce defects in a Pakistani Manufacturing Company. J Appl Eng Sci 20(2):552–561. https://doi.org/10.5937/jaes0-34558

Widz M, Krukowska R, Walas B, et al. (2022) Course of values of key performance indicators in city hotels during the COVID-19 pandemic: Poland case study. Sustainability 14(19). https://doi.org/10.3390/su141912454

Yirga AA, Melesse SF, Mwambi HG, Ayele DG (2020) Modelling CD4 counts before and after HAART for HIV infected patients in KwaZulu-Natal South Africa. Afr Health Sci 20:1546–61. https://doi.org/10.4314/AHS.V20I4.7

Zaman DM, Zerin NH (2017) Applying DMAIC Methodology to Reduce Defects of Sewing Section in RMG: A Case Study. Am J Ind Bus Manag 7:1320–1329. https://doi.org/10.4236/AJIBM.2017.712093

3 Case Study 1. Reduction of Defective Parts on an Assembly Line

3.1 INTRODUCTION

Manufacturing companies face various problems that prevent them from achieving sustainable performance. One of these is the defective parts, which can be costly (Orrantia-Daniel et al. 2019). In addition to the costs of manufacturing defective parts, other problems can arise during assembly. For example, if a defective part reaches the assembly stage and is not detected in time, it can cause failures in the performance of the final product, resulting in warranty claims by customers or even safety issues. Furthermore, the presence of defective parts affects the efficiency and performance of the entire assembly line, as operations must be stopped to correct the problem and replace the defective parts (Gao et al. 2019).

This can lead to production delays, decreased productivity, and ultimately, customer loss and damage to the company's reputation. In addition, additional costs can be generated regarding rework, waste materials, and lost time for workers who must be dedicated to correcting or replacing defective parts instead of performing other productive tasks (De Cuypere et al. 2012). Therefore, preventing defective parts is vital for maintaining efficiency and productivity in an assembly line (Hager et al. 2017).

DOI: 10.1201/9781003564607-3

To address this problem, it is crucial to implement workload-balancing strategies that balance the total operation or process time allocated to each workstation or machine on the assembly line. This approach not only optimizes the throughput of each workstation but also contributes to reducing manufacturing time and improving the overall efficiency of the assembly process (Alsaadi 2022). Additionally, it is important to implement quality control techniques at each stage of the assembly process to detect and prevent the production and passage of defective parts (Emami-Mehrgani et al. 2016).

In addition, it is essential to consider the modularity criteria in manufacturing and assembly design. Design customization and standardization of constructive components reduce the total number of defective parts, reduce manufacturing times, and improve overall efficiency. It is essential to have clear procedures for handling defective parts, which should be immediately removed from the assembly line and sent to a shop for repair or reconditioning facilities (Sabharwal et al. 2009). The necessary corrections must be made to prevent the propagation of problems in subsequent processes, which undoubtedly ensures error-free production and avoids the costs associated with the manufacture of defective parts (Bulgak and Sanders 1988).

The following is a case study of the problem of defective parts in an assembly line, the objectives, how the problem was solved by applying the DMIAC methodology and other tools, and the results.

3.2 CASE STUDY

This study was conducted at a U.S. global manufacturing service company headquartered in the gateway area of St. Petersburg, Florida. It is one of the largest companies in the Tampa Bay area, with 100 plants in 30 countries and 260,000 employees worldwide. The company was founded in 1966 and formally incorporated into Detroit in 1969. One of its first products was printed circuit boards (PCBs). Today, the divisions span numerous industries related to healthcare, life sciences, clean technology, instrumentation, defense, aerospace, automotive, computing, storage, consumer products, networking, and telecommunications. In addition, the company offers comprehensive design, manufacturing, supply chains, and product management services to customers in a wide range of industries.

The company has four manufacturing facilities in Tijuana, Mexico, which focus on the medical industry and are part of the company's regulated industries. All of these plants are ISO 13485 and Food and Drug Administration (FDA) certified. These four plants comprise more than 110,000 m^2 of manufacturing space, providing healthcare customers with the broadest range of design and manufacturing capabilities. Specifically, the present project is being conducted in Plant III, which has an area of 2,580 m^2.

The operational infrastructure of Plant III is based on the fact that it is a company that offers services for manufacturing medical products bearing the name of the same client. The company carries out approximately 30 medical device final assembly processes, with 15 final assembly lines and a subassembly area (OPEN/Lubrication). The company also has two ISO 8 and ISO 9 clean rooms and two controlled electrostatic discharge rooms.

This chapter studies the case of one customer. Figure 3.1 shows the areas where the manufacturing process begins to complete the product and each area through which the product has to pass until it reaches the shipping area. It can be observed that the process starts in the raw material receipt area, followed by the component input inspection area. The material is sent to the warehouse and then placed in the OPEN cabinets/lubrication section of Kiteo if and only if the components are applied for the material to be processed in the decorators/pad printers or require component lubrication (Dry Film, Sodium, or Parylene). The raw material was supplied in OPEN/Lubrication. Once the process is finished, the raw material is sent to the final assembly line cleanroom. However, the material is sent directly to the final assembly line if the components do not require lubrication or decoration. Once the process is completed, the material is sent to the packaging and shipping areas.

Table 3.1 summarizes each process, which can be more easily understood.

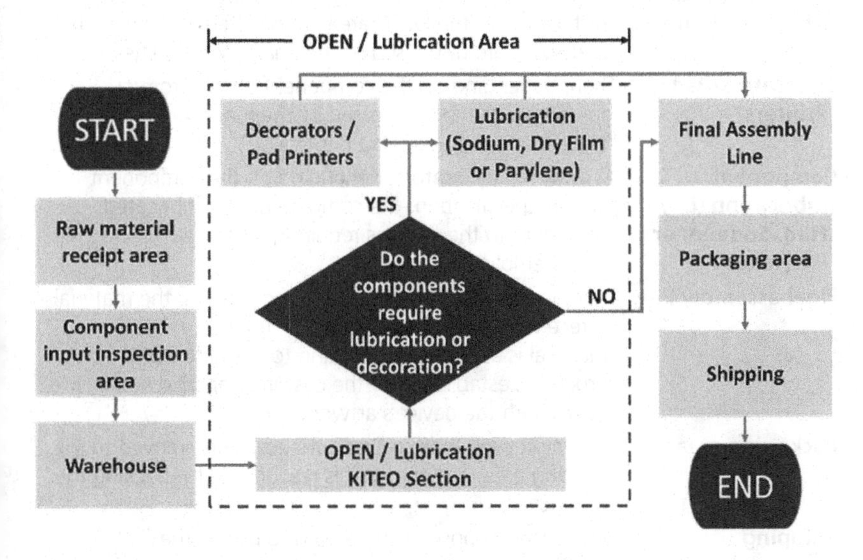

FIGURE 3.1 Flow chart of the component processing process for the case study.

Plant III comprises nine fundamental departments: Human Resources, Manufacturing, Cell Management, Quality Engineering, Finance, Health and Safety, Operational Excellence, Engineering, and Supply Chain. These departments form part of the company's work team to ensure the proper functioning of its operations.

TABLE 3.1 Description of the Processes to Complete the Product

PROCESS	*DESCRIPTION*
Raw material receipt area	In this area, the designated personnel are in charge of receiving the material from the different suppliers. They are also in charge of verifying that the documents containing the material match the destination address and that the quantities correspond to what is declared on the package containing the material. Finally, the import documents are verified.
Component input inspection area	Component input inspection is a brief inspection assigned to the component type and batch quantity. In addition, calibration documents, material certificates, and specification data are also checked.
Warehouse	Raw materials are stored in the warehouse and assigned locations according to the first-in-first-out (FIFO) method. Designated personnel are also in charge of moving the components to the different areas of KITEO.
OPEN/Lubrication KITEO Section	In the KITEO section of OPEN, the components to be distributed to the OPEN area are counted and weighed, and some are dispensed to the final assembly lines.
Decorators/Pad Printers	Then, depending on the component of the product families, these are processed by the machines of their respective decorators/Pad Printers.
Component lubrication (Dry Film, Sodium, or Parylene)	As with the decorators, depending on the component and its specification, the components are lubricated according to the families requiring lubricated subassemblies.
Final assembly line	Then, in terms of the manufacturing process, the material is received by the warehouse and lubricated. This material is assembled according to the different processes established by the customer and packed in boxes with the device's artwork.
Packaging area	The devices are then packed into boxes and moved to the packing area, and the staff is responsible for packing the products.
Shipping	Finally, the purpose of this area is to review the documentation to process the shipment and load it onto the trucks for transport to the final destination.

The following is a description of the different departments in which Plant III has to synergistically carry out all the activities of the company's general operation:

- Human Resources: This is the department in charge of the personnel recruitment process, training, professional development plan, and verification of compliance with internal procedures, from statutes related to the facilities to governmental agencies.
- Manufacturing: This department is responsible for manufacturing the products and ensuring compliance with the training required over time. It is also in charge of the metrics of adherence to the production plan.
- Cell Management: The department in charge of contacting the customer corporately. It is also responsible for managing different departments that influence the manufacturing process of the corresponding customer. It is worth mentioning that each cell manager is responsible for responding to the customer with the metrics and results of different processes.
- Quality: This department is in charge of ensuring that the processes are functional to generate a quality product according to the specifications and characteristics required by the client and comply with the regulatory systems of the medical industry. It is also responsible for ensuring that different departments of the company comply with the quality management system.
- Finance: This division manages the company's monetary resources. Its activities include making payments to suppliers and employees and approving the company's various projects.
- Health and safety: This department is responsible for ensuring the execution of worker safety plans. Its functions include industrial safety inspections, incident and accident reporting and investigation, and ensuring compliance with industrial safety training plans.
- Operational Excellence: This department is responsible for ensuring the correct progress of continuous improvement projects and implementing industrial engineering methodologies. It is also responsible for measuring the cycle times of the different processes.
- Engineering: Responsible for developing new products, new processes, changes in the documentation of work instructions, automation of processes, preventive maintenance of equipment, and generation of standards and methods to control scrap metrics.
- Supply Chain: The department in charge of production planning, material purchases and imports, inventory control, shipping, and export of manufactured products.

Plant III had approximately 3,500 employees distributed across the nine departments described previously. The organization has two schedules for the manufacturing/production department: Monday through Thursday, the morning shift from 6:00 a.m. to 6:00 p.m., and the night shift from 6:30 p.m. to 5:50 a.m. The company has two schedules for administrative staff: the night shift from 6:30 p.m. to 5:50 a.m. The company has two schedules for administrative staff: from 7:30 a.m. to 6:00 p.m. Monday through Thursday and on Fridays from 7:30 a.m. to 3:30 p.m.; the second schedule is from 6:00 a.m. to 4:00 p.m.

The process analyzed in this case study is the Pad Printing process. This process is carried out in an ISO 9 clean room, known as OPEN/Lubrication. It has eight machines, three of which are used for component lubrication. It also has four Pad Printers and a KITEO station of materials where the components are prepared for OPEN/Lubrication and the final assembly lines.

The OPEN/Lubrication area is staffed by six operations personnel, a production leader, an automation technician, a material man who supplies components to the various machines, and process engineering support.

3.2.1 Problem Statement

Figure 3.2 shows the machine in which the Pad Printing process is performed. The main features of the machine are as follows: 12 carousel-shaped nests,

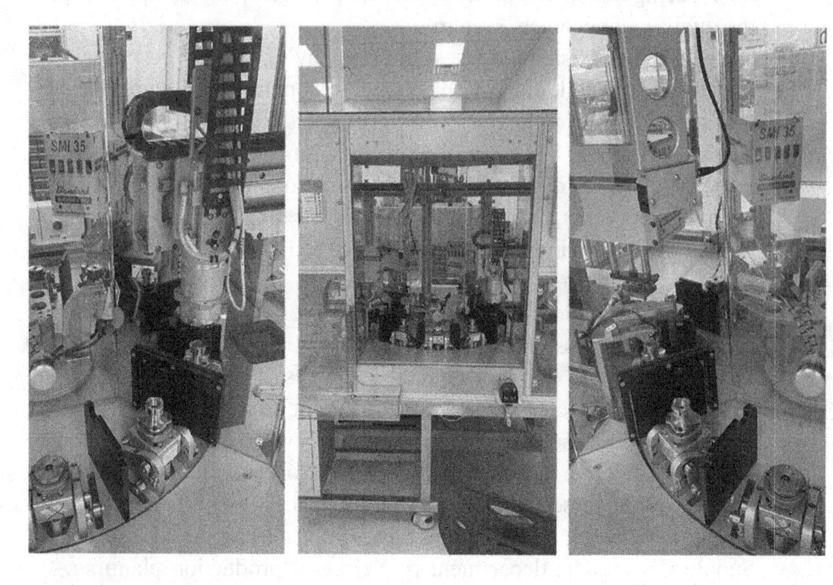

FIGURE 3.2 Pad Printing Machine.

three decorating stations, a vision system, and an electro-pneumatically controlled arm with programmable logic controllers (PLC) to deposit the defective parts in a container; otherwise, the conforming parts are deposited in the container of acceptable material, according to the vision system.

The *housing cap* decorating/Pad Printing process (see flow chart in Figure 3.3) for the production of endoscopic products starts with the arrival of raw material at the warehouse, continues with its transfer to the lubrication area, and continues with the storage of raw materials in the KITEO area in lubrication. The production order is then verified and placed in the raw material rack to be processed, as requested. The operator begins to fill out the Device Historical Record (DHR). At the start of the production order review, the automation technician starts the setup process. Once this process is complete, the master assembler releases the production order.

Subsequently, the machine operator takes the part, positions it in the nest, and then presses the button/sensor. This button/sensor detects the part as the machine carousel rotates (the operator controls all turns via the button/sensor) until the part is positioned at the decorator station. Once the part is in position, the decorator cup slides forward to imprint ink on the cliché logo. Then, the cup slides back, and the pad picks up the ink and nests where the part is located. At the same time, the pad slides forward and down and then leaves the piece decorated with the logo. The machine's carousel then rotates, and the nest is positioned on the decorator of the second logo. Then, the nest rotates to make the decoration on the other side of the Housing Cap, and the decorator

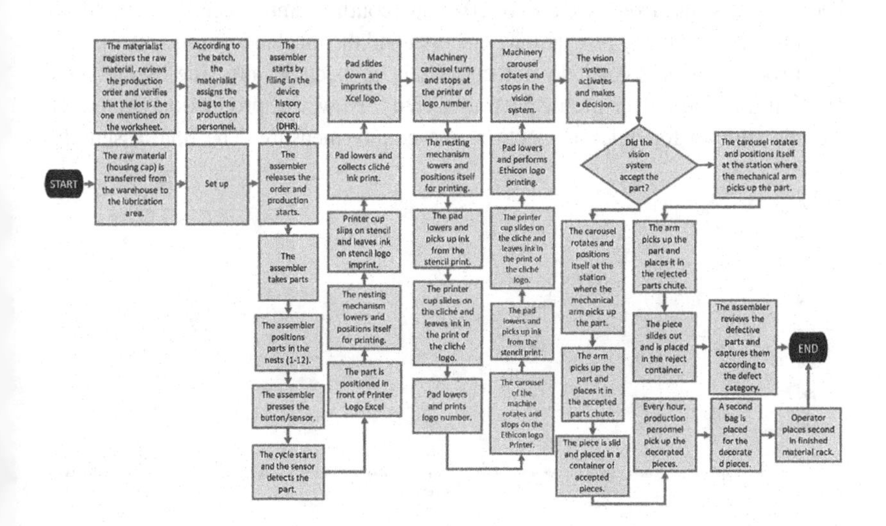

FIGURE 3.3 Pad Printing decoration flow chart.

performs the same procedure described previously. The carousel then rotates and stops until it is positioned on the decorator of a third logo. The decorator operates the cup actuating mechanism and slides the cup to the front to leave the ink on the cliché print. Next, the cup is moved back so that the pad moves down and picks up the ink, which slides to the front and moves down so the print of the first logo can be placed. Once completed, the pad goes up, and the cup returns to its original position.

The operator then pressed the button/sensor until it reached the vision system station. Next, the system took three pictures, one from each logo. In the next step, the vision system compares the current cycle photos against the master photos stored in the system's memory. The vision system was programmed to have an acceptance rate of 65%. Once the vision system has performed the analysis, it makes a decision. If accepted, the carousel rotates and positions itself at the location where the electro-pneumatic arm picks up the part and deposits it in the accepted parts chute. Otherwise, they are deposited in the rejected chute parts.

It is worth mentioning that one cycle of the Pad Printing process involves 12 pieces, that is, one piece per nest. The process was paused after one hour because it was necessary to remove the accepted material. Finally, the operator approaches the machine, takes the bag with the decorated material, places a second bag so it is not contaminated during its transfer, and places it in the finished product rack.

The project arises because the Pad Printing process is not carried out in a standardized way from the setup process to the release of the machine to start production, which means that there is a constant variation in the output of the decorated component. Figure 3.4 shows the number of defects detected over 13 weeks. The problems start from week 46 of 2021 and continue until week seven of 2022. Defects are detected by production personnel and the machine-vision system.

Production personnel do not constantly inspect because the assembler's position does not cover the inspection activity. However, production personnel provide support for quick inspection as a countermeasure. The final assembly

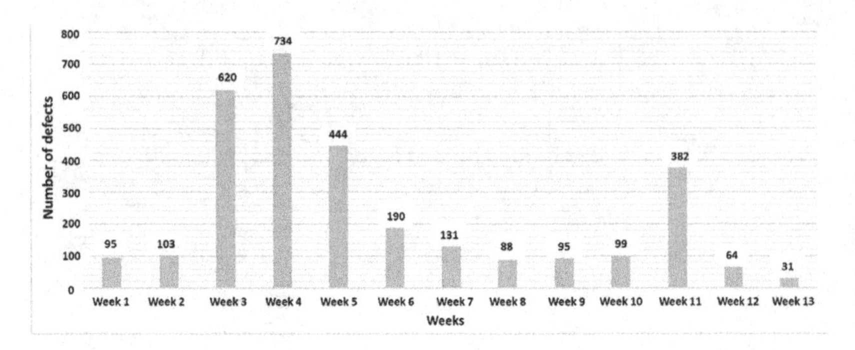

FIGURE 3.4 Number of defective parts over 13 weeks in Pad Printing.

production personnel detect the defects and reject the material. This problem is evident when the target percentage of defective parts is captured versus achieved weekly. For the Pad Printing process, the company established a weekly goal of not having more than 1% defective parts. Figure 3.5 compares the goal versus the percentage of defective parts obtained per week and shows that in 12 of the 13 weeks analyzed, the percentage of defective parts was greater than 1%.

Similarly, Figure 3.6 shows the defective parts in terms of cost (U.S. dollars, USD) against the target of 1% per week. As can be seen, there is a variation in the Pad Printing process, with weeks of defective part costs exceeding the planned budget.

Figure 3.7 presents the Pareto analysis and shows that the most recurrent defect is the incomplete Ethicon logo defect, with 1,871 defective parts. The stained Xcel logo has 292 unacceptable parts, followed by the material falling to the floor, with 126 defects. The fourth most recurrent defect is the stained logo number, with 94 parts, and finally, the watermark logo number has 93 rejected parts. These five defects are the major contributors to unacceptable parts in Xcel Universal's Pad Printing decorating process.

Finally, it is necessary to mention that the Pad Printer process presents deficiencies in the downtime metric, shown in Figure 3.8, expressed in hours, recorded during each of the 13 weeks analyzed and compared with the

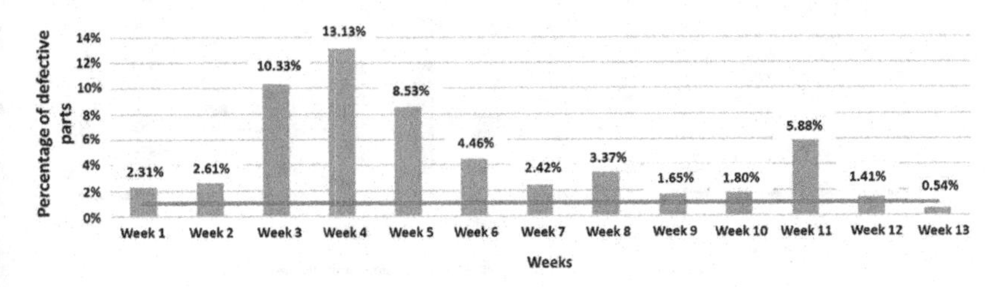

FIGURE 3.5 Percentage of defective parts in the Pad Printing process.

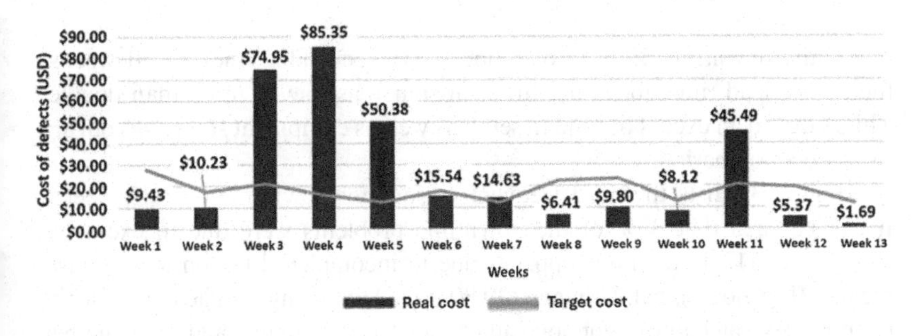

FIGURE 3.6 Comparison of actual defect costs versus target costs.

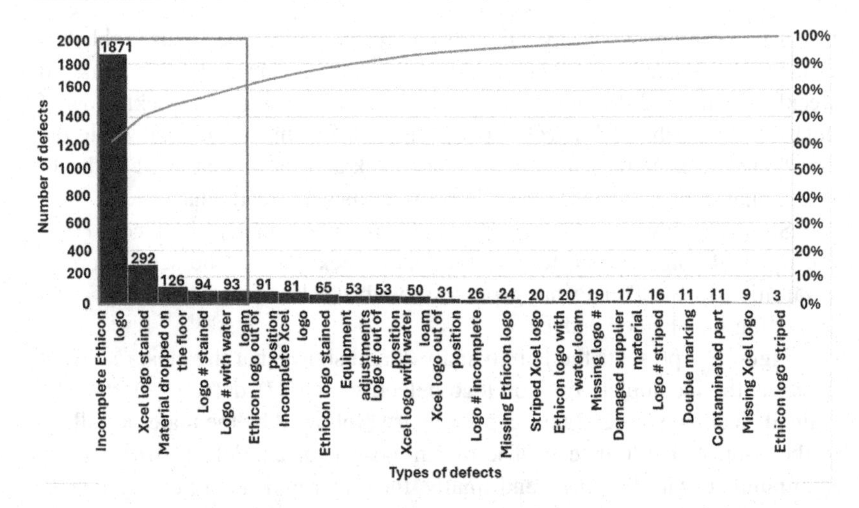

FIGURE 3.7 Pareto diagram of the types of defects in the Pad Printing process.

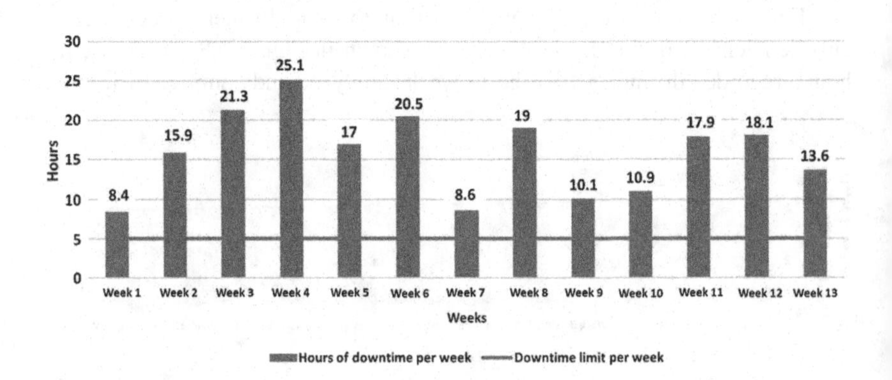

FIGURE 3.8 Dead time of the Pad Print process.

maximum downtime allowed of 5 hours per week (blue line). In all weeks, there was dead time above the limit, in some cases even more than double. This is due to an excess of time in setup as well as equipment stoppages during the production period.

Figure 3.9 lists the most frequent reasons for downtime over 13 weeks. It can be seen that 80% of the downtime problems were due to excessive setup time (112.47 hours), stoppages due to incomplete Ethicon logo adjustments (21.78 hours), paint change (20.80 hours), cleaning on Pads and clichés (17 hours), scratched Ethicon logo adjustments (11.13 hours), and logo number adjustment with watermark (9.40 hours).

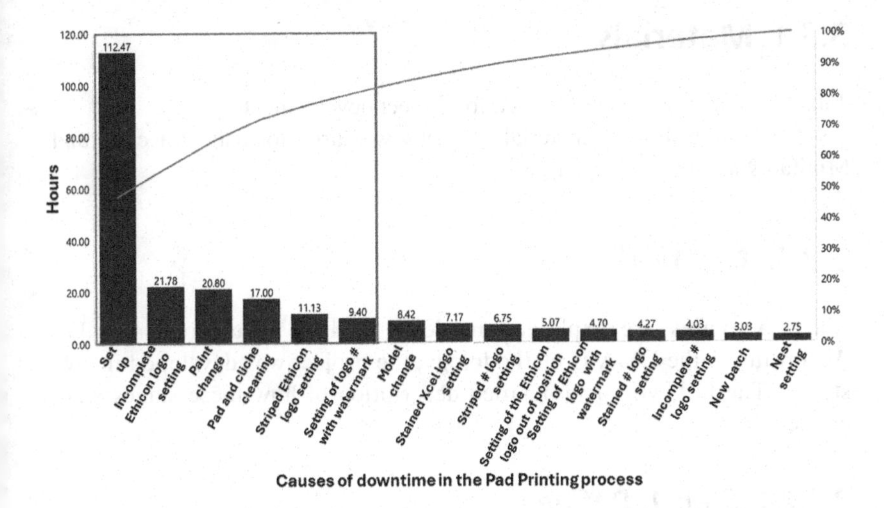

FIGURE 3.9 Pareto diagram of causes of downtime in the Pad Printing process.

It is concluded that the DMAIC methodology and auxiliary tools are helpful for companies to eliminate defects and their causes, reduce downtime and active time of production processes, and eliminate other waste (transportation, unnecessary operations, and delays). Consequently, companies reduce costs and improve their competitiveness in the market.

3.2.2 Objectives

The general objective of this project is to reduce defects in the Pad Printing process using the DMAIC methodology; however, there are the following specific objectives:

a. Reduce Pad Printing process downtime to less than 5 hours per week.
b. Reduce the setup time of Pad Printing by 50%.

3.3 MATERIALS AND METHODS

This section presents the methodology for conducting the project and achieving the proposed objectives. This section is divided into two main subsections: Materials and Methods.

3.3.1 Materials

Table 3.2 shows the materials used for project development.

In addition to these materials, Excel® was used for data collection and Minitab® for statistical analysis.

3.3.2 Method

These activities are explained in detail and in chronological order. According to the DMAIC methodology, the project was developed in five stages. The following is a detailed description of how these stages were carried out.

3.3.2.1 Step 1. Define

In this stage, the problem to be solved is defined. To do so, planning involves the expectations and needs of customers. Questions such as: What causes the defects in Xcel Universal's Pad Printing equipment, the scope of the defective parts concerning the cost, and the cost of the defective parts are raised and answered. The answers are acquired through documentation and analysis of the pad printing process through a supplier, inputs, process, outputs, and customers (SIPOC) analysis, a Pareto diagram, and a project charter. The following is a description of each analysis activity in this phase.

- **SIPOC analysis:** Defining the suppliers, identifying the inputs, process itself, outputs, and customer requirements.
- **Pareto Diagram:** An analysis is made with the Pareto diagram to define the problems or circumstances that contribute to the general problem of the process. Thus, it is defined as 20% of the defects generating 80% of the general problem.
- **Project charter:** This is elaborated upon by considering the elements proposed by Oviedo (2016). The project charter includes project objectives, justification, a high-level description of the project, high-level requirements, success criteria, high-level risks, a summarized schedule of activities, a summarized budget, and a list of stakeholders.

In addition, the objectives to be achieved are considered, the tools to be used are determined, the process and its interactions are identified, and the critical variables within the process are identified.

TABLE 3.2 Materials Used for Project Development

MATERIAL	*QUANTITY*	*IMAGE*
Housing Caps	3776 pcs.	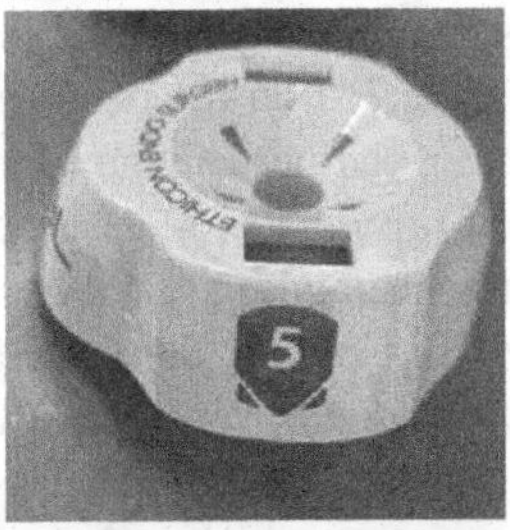
Pad Printer Equipment	1	
Ink cups	3 pieces	
New clichés	3 pieces	
Speed controllers with numbering	12 pieces	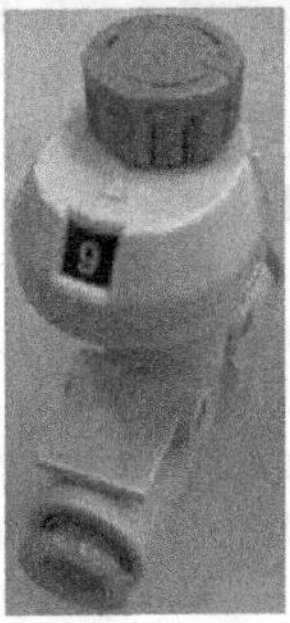

3.3.2.2 Step 2. Measure

This stage begins with collecting data and measuring the number of defective parts, machine downtime, speed regulator levels, and ink viscosity. Data were also collected on the dead times at each stage of the process, the percentage of scrap, the percentage of stoppages, and quality rejections, among others. In addition, the product acceptance criteria are clarified based on customer documents, and it is established how defects and downtime will be measured. Likewise, how the data will be taken and the strategy to collect the information are defined. What, how, when, and who should perform the measurements is established through check sheets, research, and registration to use descriptive statistics, which include data collection, tabulation, analysis, and interpretation of quantitative data and, in this way, to make decisions.

In this phase, check sheets are used to capture defects with different categories and codes. The ink viscosity and values assigned to the machine speed controllers (pad back, pad front, pad up, pad down) were also recorded. On the other hand, the log sheet is also used to notify the automation technician of defects hourly to make equipment adjustments. Finally, the information was recorded in the database using Excel® software to synthesize the information through graphs.

3.3.2.3 Step 3. Analyze

At this stage, the main factors affecting the operation of the process are determined based on the information gathered in the previous phase. To this end, statistical tools and analysis methods have been applied, such as brainstorming, flow diagrams, Pareto diagrams, and Ishikawa diagrams. In addition, graphs were constructed using the data obtained in the previous phase.

- **Brainstorming:** This activity requires meeting the entire Pad Printing process team. Each member is asked for their opinion to define the factors that affect or generate the Pad Printing problem.
- **Flow chart and process flow diagram:** The setup of the machine is presented. In addition, the activities that give rise to defects in the Housing Caps processed at Pad Printing are identified.
- **Ishikawa Diagram:** From the information collected through brainstorming, a grouping is generated according to the factors of the 6M method: labor, machinery, measurement, method, environment, and materials (Morales 2007).
- **Construction of relationship graphs:** The collected data are arranged graphically to define cause-and-consequence relationships, thus concisely exposing the factors.

3.3.2.4 Step 4. Improve

Different solutions are exposed in this stage by identifying the relevant variables and the appropriate levels at which the process should operate. Statistical and lean manufacturing tools were used to perform these activities. Improvement ideas are captured, solution alternatives are evaluated and selected, suggestions from the work team and top management are presented, and changes are implemented. This was carried out through the following steps:

a. **Purchase of clichés and cups for Pad Printing:** This is done to reduce defects due to cliché wear and to have spare cups to reduce setup times. In stage 3 of "Analyze," it was determined that a possible effect on the defects of the Housing Caps is cliché wear. Therefore, we decided to fix this variable and proceed with purchasing. To do so, the equipment supplier is contacted, and new clichés and cups are requested. Once the material arrives at the plant, the automation technician performs the change in equipment parts, and tests are carried out for a week to verify that the equipment is working correctly.
b. **Installation of speed controllers with numbering:** To carry out this activity, the equipment printers change the speed controllers. This allowed the speed of the printing system to be controlled. The speed is controlled by four inputs that lead to the *pad's back, front, up, and down* outputs.
c. **Definition of factors and their levels:** In this step, the ink speed and viscosity factors are considered, and the trial-and-error method is applied. First, the ink was prepared at different viscosity levels, and adjustments were made to the regulators without altering the validated state or affecting the equipment operation. Subsequently, high and low values of the speed and viscosity of the ink were determined.
d. **Planning the Design of Experiments (DOE):** The design of experiments (DOE) was carried out in Minitab with personnel from the Quality Department. The experiment evaluated five factors: speed regulators (*Pad Back, Pad Front, Pad Up, Pad Down*) and ink viscosity. The evaluation was conducted with 59 pieces per run, as established in the company's quality system attribute evaluation procedure. As a five-factor DOE with a full factorial design, two replicates and 32 runs per replicate were required for 64 runs. The number of pieces evaluated was 3,776. To carry out DOE planning and evaluate the feasibility of the design execution and the availability of production time and materials, the production, engineering,

planning, and quality teams are convened to meet. Subsequently, the quality and engineering team develops a "Pass/Fail" inspection sheet to verify run parts. The output of the DOE run is the percentage of accepted parts per run, and based on the result, the DOE was run using the Minitab® software.

e. **DOE execution:** The design execution consists of presenting the order of the design runs to the lubrication team and client and defining the number of parts to be evaluated. In addition, the document from which the reference of the acceptance or rejection criteria is taken is defined, as well as the values of the levels of the five factors. The following describes how DOE was performed:
   - A lubrication materialist is required to supply the material to the equipment.
   - The production staff is trained in the engineering study, which states that every 59 decorated parts must be removed from the equipment and evaluated by the inspector and quality engineer of the final assembly line. In addition, the automation technician is notified when adjustments are made according to the run number.
   - The automation technician adjusts different factors and prepares the ink for both levels (low and high).
   - Manufacturing engineers and automation technicians segregate and label materials based on the run number.
   - The quality engineer and inspector filled out the form to record the accepted and rejected parts.
   - Once the 3,776 parts have been decorated, the manufacturing and quality engineers record the quantities of accepted and rejected parts and place the data in the Minitab® generated design defined in the previous step.

f. **Use of single-minute exchange of dies (SMED) lean manufacturing tool to reduce the setup time:** To reduce dead time, the levels at which the equipment works properly are defined, and through the execution of the DOE, the values of speed and viscosity at which an acceptable output of decorated material can be obtained are defined. Therefore, the SMED technique was applied, which consisted of the following four fundamental stages (Espin 2013):
   - Observe and understand the process.
   - Identify and separate operations.
   - Convert internal to external operations.
   - Standardization of the new method.

g. **Complete implementation of the activities:** Once the statistical analysis is performed with the DOE, the factors that impact the decorating process are obtained, and decisions are made to improve the output of the decorating process. In addition, the purchase of ink cups for the Pad Printer is carried out to eliminate tasks in the setup process. In addition, the new setup method is established, making a new process flow diagram and the changes in the corresponding documentation, in which the setup method is redefined, and the parameters in which the printer speeds and ink viscosity must be found are placed. Finally, the personnel involved in the new Pad Printer process were trained.

3.3.2.5 Step 5. Control

At this stage, the effectiveness and efficiency of the various changes undergone by the process were tested. Indicators that show the level of performance of the actions are displayed, and the status of the process before the start of the project versus after its completion is compared. In addition, appropriate documentation for the changes made was implemented.

In this stage, the implementation of the improvements from the previous phase was evaluated using scrap and dead time indicators. According to the evaluation results, control actions were presented to strengthen the improvements, such as updating the instruction sheets of the Xcel Universal pad Printing process and creating registration forms to ensure compliance with the parameters. The results were presented at a meeting with the work team of the work cell.

3.4 RESULTS

This section presents the results obtained using the method described previously. For better understanding, these results are presented for each stage of the method.

3.4.1 Results of Step 1: Define

This section presents the results of the activities conducted during the definition stage.

3.4.1.1 Elaborate Project Charter

The result of this activity was to establish the project title, define the project leader and process owner, locate the case study, manage the project, and describe the business case. Subsequently, the problem was defined, the project's scope was established, and the objectives and metrics that would be impacted were determined. The main actions of the project included commitment dates and those responsible for the activities. Finally, the required support and possible risks the project could present during its execution were included. The result of the case study description corresponds to what is mentioned in Section 3.2, and the problem statement is mentioned in Section 3.2.1, while the objectives are mentioned in Section 3.2.2.

3.4.1.2 Performing SIPOC Analysis

Table 3.3 shows the result of the SIPOC analysis. The process elements, such as suppliers, inputs, processes, outputs, and customers, were identified. The inputs marked in red are significantly related to the process. The customer's expectation regarding the output of Xcel Universal's Pad Printing process was also defined. This analysis identifies the factors influencing the process and definition of suppliers.

3.4.1.3 Performing the Pareto Analysis

The results of the Pareto diagram for the defects generated by the Pad Printer machine for 13 weeks are shown in Figure 3.7. It was found that 80% of the problems were composed of incomplete Ethicon logo with 1871 defective pieces, Xcel logo stained with 292 defective pieces, material dropped on the floor with 126 defective pieces, logo number stained with 94 defective pieces, and logo number stained with watermark with 93 defective pieces.

The Pareto diagram for downtime in the Pad Printer process is shown in Figure 3.9. It can be seen that 80% of the downtime problems were due to excess time in setup at 112.47 hours, stoppages due to incomplete Ethicon logo adjustments at 21.78 hours, paint change with 20.80 hours, cleaning of *Pads* and clichés with 17 hours, scratched Ethicon logo adjustments with 11.13 hours, and adjustment of logo number with watermark with 9.40 hours.

3.4.2 Results of Step 2: Measure

This section presents the results of the activities carried out in stage two of the DMAIC methodology.

Table 3.4 shows the results of the speed level combinations. Four configurations were defined with speed levels at which the machine did not present irregularities in its operation and fulfilled the objective of pouring the ink into the cliché print.

TABLE 3.3 SIPOC Analysis of the Pad Printing Process

SUPPLIERS	*TICKETS*	*PROCESS*	*EXITS*	*CUSTOMERS*
Providers of the necessary resources	Resources required for the process	Description of the activity: Requirements	Process deliverables	Final Assembly Line: Requirements
Production Department	Labor	Trained personnel	Personnel perform the activity according to the method described in the documentation.	The component decoration is legible, unblemished, and complete.
Engineering Department	Setup	Trained personnel	Personnel perform the activity according to the method described in the documentation.	
Automation Department		Equipment and tools are in good condition.	The setup of the machine is carried out in the shortest possible time.	
Engineering Department		Ink preparation should be done using the recommended viscosity.	The ink has adequate viscosity, and the decoration is legible, unblemished, and complete.	
Warehouse	Ink	The chemical must meet the requirements of the safety data sheet.	The chemical complies with the requirements of the safety data sheet.	
Warehouse		Must be used before the expiration date.	The chemical complies with the requirements of the safety data sheet.	
Warehouse	Housing Cap	Molded parts must conform to drawing specifications.	Molded parts comply with drawing specifications.	

(Continued)

TABLE 3.3 (Continued)

SUPPLIERS	*TICKETS*	*PROCESS*	*EXITS*	*CUSTOMERS*
Warehouse	Thinner	The chemical must meet the requirements of the safety data sheet.	The chemical complies with the requirements of the safety data sheet.	
Warehouse		Must be used before the expiration date.	The chemical complies with the requirements of the safety data sheet.	
Maintenance Department	Machine	According to its validated status, the main pressure gauge should be around 80 ± 10 pounds per square inch (PSI).	The central pressure gauge complies with the specification of 80 ± 10 PSI, according to its validated status.	
Automation Department		Air pressure and velocity of Pad Back, Pad Front, Pad Up & Pad Down 0%-100%.	The decoration of the component is legible, unblemished and complete.	
Automation Department		All three printing stations must have clean and unworn pads and clichés.	The component decoration is legible, unblemished, and complete.	
Automation Department		The nests where the component is placed must be aligned.	The component decor is legible, unblemished and complete.	

TABLE 3.4 Pad Printer Speed Level Combination Settings

CONFIGURATION	*PAD BACK*	*PAD FRONT*	*PAD UP*	*PAD DOWN*
1	6	6	8	8
2	8	8	6	6
3	6	8	6	8
4	8	6	8	6

Regarding the equipment speed levels, ink viscosity, and defects, ten samples were taken from the viscosity level, each with 24 data, as six times were obtained for each setting level. The results indicated that, as the viscosity decreased, fewer defective parts were obtained. The lowest defects (5 and 6) occurred at an average viscosity level of 2.39 and 2.42, respectively, in configuration 1. On the other hand, the highest amounts of defects (32 and 39) occurred when the viscosity levels were 3.51 and 3.44, respectively, both in configuration 2.

3.4.3 Results of Step 3: Analyze

The results of the activities conducted during the analysis stage are presented in what follows:

- **Elaborate flow chart and process flow diagram:** Figure 3.3 shows the flow diagram of the current Pad Printing decorating process. Table 3.5 shows a summary of the initial process flow diagram of the Pad Printing Setup. The process consisted of 144 activities, totaling 47.8 minutes and 287.1 meters of transport.
- **An Ishikawa diagram is constructed, as shown in** Figure 3.10. The incomplete Ethicon logo defect represented the problem because, in the Pareto diagram, it was the major contributor to equipment defects. The causes evaluated during the measurement stage of the DMAIC cycle are associated. The numbers before each cause represent the priority levels assigned to each cause.
- **Performing graphs of defect ratio, viscosity, and speed settings on Pad Printer:** When performing the graphs, it was found that there were more defects when there was a combination of ink viscosity between 3.40 and 3.50 seconds and a speed setting of 2. Similarly, minor defects occurred when configuring 1 (the speed levels of Pad Back, Pad Front, Pad Up, and Pad Down were 6, 6, 8, and 8, respectively) and viscosity of 2.35 and 2.50 seconds.

TABLE 3.5 Summary of the Current Pad Printing Setup Process Flow Diagram

VARIABLES	*OPERATION*	*TRANS-PORTATION*	*WARE-HOUSE*	*DELAY*	*INSPECTION*	*TOTAL*
Time (sec)	2292	272	0	30	273	2867
Distance (m)	0	287.81	0	0	0	287.81

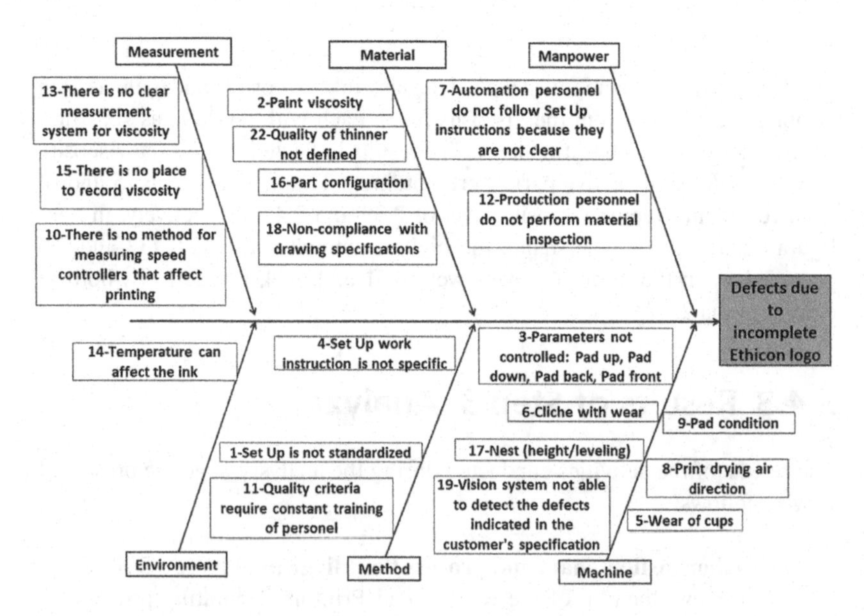

FIGURE 3.10 Ishikawa diagram for incomplete Ethicon logo defects.

3.4.4 Results of Step 4: Improve

The results of the activities conducted during the analysis stage are presented in what follows:

- **New wear-free clichés were purchased.** Thus, the cliché depth was the same in all areas and in accordance with the customer's drawing specifications. In addition, the purchase of ink cups facilitated the ink change at the beginning of each shift, eliminating cleaning activities during setup at the beginning of the shift.
- **Numerical speed controllers are also installed.** As a result of this improvement, there was more significant control over the speed

factors of the equipment. These factors, Pad Back, Pad Front, Pad Up, and Pad Down, are fundamental to performing the DOE, which contributes to defining at which levels the equipment works properly, reducing the production of defective parts.

- **Run DOE:** Sixty-four runs were performed, and the ink viscosity was of two levels: 2.5 and 3.5. The percentage of accepted parts was between 53% and 100% and was higher at a viscosity level of 2.5. The ANOVA indicated that viscosity had the most significant impact on the quality of the pieces. In addition, Pad Back speed, the combination of Pad Back and Pad Down speed, and Pad Down and Pad Front were significant factors since a p-value <0.05 was obtained, and the R^2 value was 88.51%. DOE suggested the speed configuration of 6, 6, 7, and 8 for Pad Back, Pad Front, Pad Up, and Pad Down, respectively, and a viscosity of 2.50 seconds as the optimal combination.
- **Apply the SMED tool for setup-time reduction,** which allows for identifying and separating internal and external operations, which are then converted to internal operations. With these changes, 44 activities, 1323 seconds, and 77.94 meters were required to carry out the Pad Printing setup process, representing a decrease of 69%, 54%, and 73%, respectively. Table 3.6 shows a flow chart summary of the new process method.

3.4.5 Results of Step 5: Control

The results of the activities conducted during the control stage are presented in what follows:

- **Update setup instruction:** Figure 3.11 shows a sample of the work instruction sheet to perform the machine's setup. The ink preparation method was established with optimal formulation quantities and viscosity for the DOE process. Simultaneously, it was established that cleaning activities would always be performed at the end of the shift. The speed levels at which the equipment should operate are also established according to the results of the optimized DOE model.
- **Present defect indicators:** Figure 3.12 presents a graph of the percentage of defects during the 11 weeks after implementing the changes. In all weeks, the percentage of defective parts was below 1%, with 0.66% being the highest percentage, averaging 0.45% per week compared with 4.5% for the 13 weeks before implementing

TABLE 3.6 Summary of the Flow Chart of the New Pad Printing Setup Process

VARIABLES	*OPERATION*	*TRANS-PORTATION*	*WARE-HOUSE*	*DELAY*	*INSPECTION*	*TOTAL*
Time (sec)	800	125	0	0	398	1323
Distance (m)	0	77.94	0	0	0	77.94

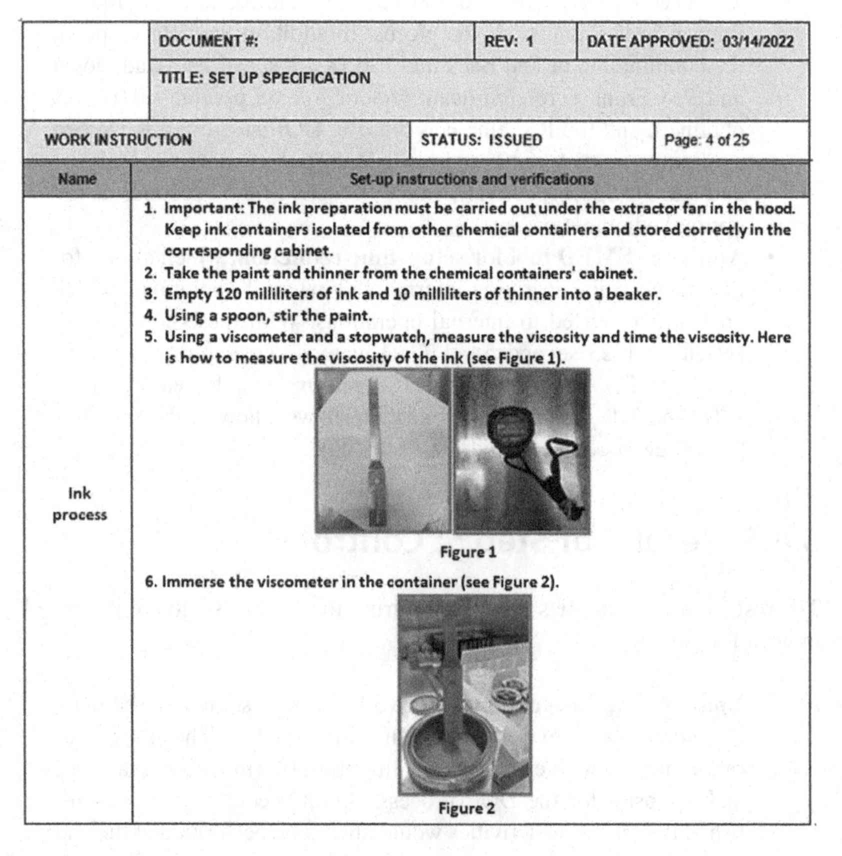

	DOCUMENT #:	REV: 1	DATE APPROVED: 03/14/2022
	TITLE: SET UP SPECIFICATION		
WORK INSTRUCTION	STATUS: ISSUED	Page: 4 of 25	

Name	Set-up instructions and verifications
Ink process	1. Important: The ink preparation must be carried out under the extractor fan in the hood. Keep ink containers isolated from other chemical containers and stored correctly in the corresponding cabinet. 2. Take the paint and thinner from the chemical containers' cabinet. 3. Empty 120 milliliters of ink and 10 milliliters of thinner into a beaker. 4. Using a spoon, stir the paint. 5. Using a viscometer and a stopwatch, measure the viscosity and time the viscosity. Here is how to measure the viscosity of the ink (see Figure 1). Figure 1 6. Immerse the viscometer in the container (see Figure 2). Figure 2

FIGURE 3.11 Pad Printer setup instruction.

improvements. In addition, 3,076 defective parts were initially recorded, compared to 1,005 after the project was implemented, representing a reduction of 67.33%. In the specific case of defective parts due to incomplete Ethicon logo, these were reduced from 1,871 to 33, representing a 98.24% reduction.

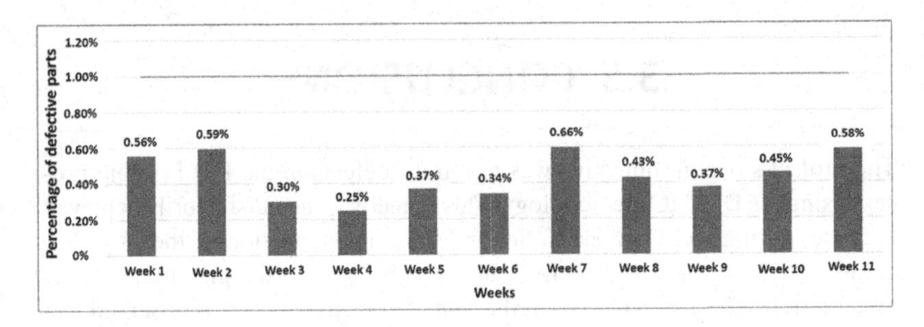

FIGURE 3.12 Percentage of defective Pad Printing parts after implementing the improvements.

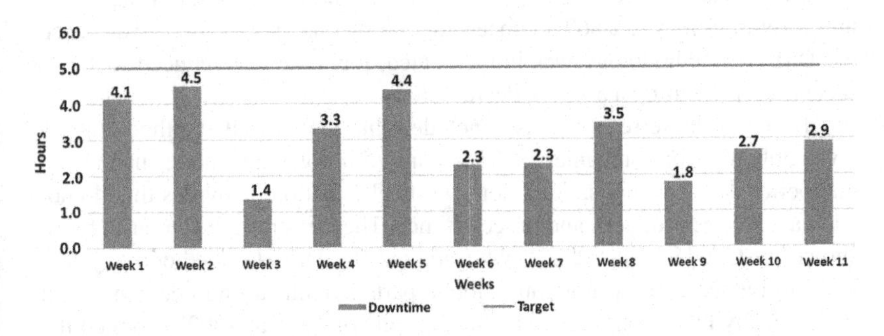

FIGURE 3.13 Downtime of the Pad Printing process after implementing the improvements.

- **Show downtime indicators:** Figure 3.13 shows the downtime of the Pad Printing process during the 11 weeks after project implementation. In all weeks, the downtime was less than 5 hours, which means that the goal was met. However, even after implementing the improvements, the task that generated the most downtime was setup, reduced from 112.7 to only 21.92 hours. This was followed by 2.02 hours for adjusting the striped Ethicon logo, 1.67 hours for adjusting the logo number with watermark, 1.05 hours for cleaning the pad and clichés, and 0.97 hours for adjusting the striped logo number. Some activities, such as incomplete Xcel logo adjustment, incomplete Ethicon logo adjustment, and Excel logo adjustment with watermarks, were reduced to zero hours of downtime.

3.5 CONCLUSIONS

The project's overall objective was to reduce defects in the Pad Printing process using the DMAIC methodology. This objective was satisfactorily achieved because, during the 11 weeks following the implementation of the improvements, the percentage of defective parts was below 1%, the maximum allowable limit. In addition, a 67.33% reduction in defective parts was observed. On the other hand, the specific objectives were to reduce the dead time of the Pad Printing process to less than 5 hours per week and to reduce the setup time by 50%. Regarding the first specific objective, it is concluded that this was achieved satisfactorily since, during the 11 weeks following the project, the maximum dead time recorded was 4.5 hours in the second week. Finally, the initial setup time was 2,867 s concerning the second specific objective. Once the DMAIC methodology was implemented, this time was reduced to 1,323 seconds, representing a 53.85% decrease.

Based on these results, it is concluded that the DMAIC methodology is a valuable tool for companies to significantly improve indicators, such as the number of defective parts, dead time, waste elimination (activities that do not add value to the product), and process times. These results are similar to those shown by Condé et al. (2023), who applied the DMAIC methodology and DOE to reduce defects in an automotive parts manufacturing company from a chronically high to an acceptable level. Zulkarnaen et al. (2023) applied the DMAIC methodology to reduce automotive paint defects. Finally, in the third case, Montororing et al. (2022) applied the same tool to reduce the defects in medical products from 2.26% to 0.93%. In all three cases, the causes of the defects were detected, and the capability of the analyzed process was improved.

It can be concluded that the DMAIC methodology, together with auxiliary tools, is helpful for companies to eliminate defects and their causes, reduce downtime and active time of production processes, and eliminate other waste (transportation, unnecessary operations, and delays). Consequently, companies reduce costs and improve their competitiveness in the market.

REFERENCES

Alsaadi N (2022) Assessment and enhancement of the manufacturing productivity through discrete event simulation. IOP Conf Ser Mater Sci Eng 1222:012011. https://doi.org/10.1088/1757-899X/1222/1/012011

Bulgak AA, Sanders JL (1988) Integrating a modified simulated annealing algorithm with the simulation of a manufacturing system to optimize buffer sizes in automatic assembly systems. In: Abrams MA, Haigh PL, Comfor JC (eds) Winter Simulation Conference Proceedings. Association for Computing Machinery, pp. 684–690.

Condé GCP, Oprime PC, Pimenta ML, et al. (2023) Defect reduction using DMAIC and Lean Six Sigma: A case study in a manufacturing car parts supplier. Int J Qual Reliab Manag 40:2184–2204. https://doi.org/10.1108/IJQRM-05-2022-0157/FULL/XML

De Cuypere E, De Turck K, Fiems D (2012) Performance analysis of a decoupling stock in a make-to-order system. IFAC Proc 45:1493–1498. https://doi.org/10.3182/20120523-3-RO-2023.00177

Emami-Mehrgani B, Neumann WP, Nadeau S, et al. (2016) Considering human error in optimizing production and corrective and preventive maintenance policies for manufacturing systems. Appl Math Model 40:2056–2074. https://doi.org/10.1016/J.APM.2015.08.013

Espin F (2013) Técnica SMED. Reducción del tiempo preparación. 3 Ciencias, pp. 1–11.

Gao Y, He C, Zheng B, et al. (2019) Quantifying the complexity of subassemblies in a fully automated assembly system. Assem Autom 39:803–812. https://doi.org/10.1108/AA-09-2018-0145/FULL/XML

Hager T, Wafik H, Faouzi M (2017) Manufacturing system design based on axiomatic design: Case of assembly line. J Ind Eng Manag 10:111–139. https://doi.org/10.3926/jiem.728

Montororing YDR, Widyantoro M, Muhazir A (2022) Production process improvements to minimize product defects using DMAIC six sigma statistical tool and FMEA at PT KAEF. J Phys Conf Ser 2157:012032. https://doi.org/10.1088/1742-6596/2157/1/012032

Morales J (2007) Aplicación de la metodología seis sigma, en la mejora del desempeño en el consumo de combustible de un vehículo en las condiciones de uso del mismo. Universidad Iberoamericana.

Orrantia-Daniel G, Sánchez-Leal J, de la Riva-Rodríguez J, et al. (2019) Predicción del Número de Paros de Producción en Líneas de Ensamble. Epistemus 13:29–35. https://doi.org/10.36790/EPISTEMUS.V13I26.93

Oviedo L (2016) Modelo de iniciación y planeación de proyectos para la empresa ChemTreat. Universidad Industrial de Santander.

Sabharwal A, Syal M, Hastak M (2009) Impact of manufactured housing component assembly redesign on facility layout and production process. Constr Innov 9:58–71. https://doi.org/10.1108/14714170910931543/FULL/XML

Zulkarnaen I, Kurnia H, Saing B, et al. (2023) Reduced painting defects in the 4-wheeled vehicle industry on product type H-1 using the Lean Six Sigma-DMAIC approach. J Sist Dan Manaj Ind 7:179–192. https://doi.org/10.30656/JSMI.V7I2.7512

Case Study 2. Production Increase

4

4.1 INTRODUCTION

Manufacturing firms face several challenges associated with low output (Harvey et al. 1992). These problems include inefficient processes, obsolete technology, inadequate training of the workforce, and supply chain disruptions (Clegg et al. 2002). To address these problems effectively, companies must analyze their current operations and identify areas for improvement. By applying lean manufacturing principles, investing in advanced technology, and continuous employee training, companies can optimize their production processes and achieve higher levels of efficiency. In addition, building solid relationships with reliable suppliers and implementing robust quality control measures can help mitigate disruptions and ensure consistent production (Herron and Braiden 2006).

Addressing the problem of low production requires a multifaceted approach that considers the various aspects of the manufacturing process. Manufacturing companies must also consider the role of human factors in low production and how employee commitment, motivation, and satisfaction can significantly influence productivity (Herron and Braiden 2006). Thus, creating a positive work environment, encouraging teamwork, and providing opportunities for professional growth can improve overall productivity. In addition, implementing performance incentives and recognition programs can encourage employees to strive for higher productivity levels (Boyle and Scherrer-Rathje 2009).

Recently, during the COVID-19 pandemic, manufacturing companies have faced unprecedented challenges, such as low production in the face of increasing demand (Kapoor et al. 2021). This situation has created difficulties in keeping

DOI: 10.1201/9781003564607-4

up with market expectations, which, in turn, has put significant pressure on the supply chain and its ability to meet consumer needs (Cai and Luo 2020). Low production can be related to several factors, such as raw material shortages, disruption of operations due to health and safety protocols, and decreased labor capacity due to illness and mobility restrictions (Okorie et al. 2020).

These challenges have led to delays in manufacturing, making it difficult to fulfill outstanding orders and maintain an adequate inventory level. In addition, manufacturing companies had to adapt quickly to new working methods, implementing telecommuting in jobs that had not previously allowed it (Okorie et al. 2020). The rapid implementation of health and safety protocols and the need to train personnel in new practices have also impacted the efficiency and productivity of operations (Kapoor et al. 2021).

On the other hand, high demand has been driven by several factors, such as the need for health- and safety-related products, adaptation of business operations to the digital environment, and changing customer consumption patterns (Armijal et al. 2023). This situation created an imbalance between supply and demand, increasing pressure on manufacturing companies (Taqi et al. 2020). As these companies face this challenge, they must develop sound strategies to overcome low production and meet the high demand. This may include diversifying sources of raw materials, implementing workplace safety measures to protect employees, and optimizing supply chains to ensure the efficient distribution of products (Kumar et al. 2020).

The following is a case study on the problem of low production in a manufacturing company. It sets objectives and explains how the problem was solved using the DMAIC methodology and other tools. The results obtained are also included.

4.2 CASE STUDY

The present study was conducted at a U.S.-based company that manufactures personal protective equipment and other equipment for law enforcement, public safety, military, and recreational markets. The armor manufactured by this company has protected more than 2,040 police officers who have been shot in the line of duty.

The company's operational infrastructure falls into the following three categories:

- Manual operations
- Automatic operations
- Semi-automatic operations

Manual operations correspond to all the work performed by workers with textiles, such as the sewing of a piece, assembly of accessories and complements of the piece, and all the manual work that influences the output of the process. Automatic operations are those that already have a programmed or predetermined pattern or software that processes a piece in an automated manner; for which there are machines of all types, including CNC cutting and sewing machines, pneumatic presses, riveting machines, laser cutting machines, thermoforming presses, and heat machines. Finally, semi-automatic operations correspond to those that involve both manual and automatic operations. An example is the processing of a belt, which involves a cutting process in a CNC machine, then in the sewing area, the final finishing, and finally, in a press, the buckles are placed to finish the piece.

The human resources, purchasing, finance, security, maintenance, planning, quality, and production departments are within the organizational structure. The last department is divided into four areas, which are responsible for the production of all services offered by the company. The areas in the production department are:

Body Armor: This is the area where the project is focused and is divided into sections, starting with the CAD-CAM section, where the marks and drawing patterns that indicate the figure and size of the vest or accessory to be processed are made, which then go to the cutting section. In this section, the shape of the vest is detected using a pre-programmed cutting machine, and all the layers of fabric that need to be processed are cut. In this section is a laser cutting machine, the details of which are made of neoprene fabric included in some vests. Then, the process moves to the subassembly section, where some accessories included in the vest are added, logos are placed, and the automated process of placing rivets on the models is performed in specific models.

All the materials needed for the sewing process, such as Velcro, Webless, and Plastic, were also controlled. Finally, the process proceeded to the sewing section. In the last section, the final assembly process is carried out by adding all the parts that make up the vest. The inspection and quality review station is at the end of the sewing section, where the station manager evaluates the complete vest process and records the quantity, order number, and total efficiency at the end of the line.

Duty Gear: In this area, molding, assembly, and sewing processes of all types of sets for the armament of police officers are performed. It begins with the cutting area, which consists of CNC machines programmed to cut the piece with the respective material to be used for its construction as well as the shape of the accessory to be made, such as holsters, retainers, keepers, pouches, or harnesses. Afterward, the assembly or sewing area is used for molding, construction, and final finishing of the piece.

Bianchi: The process is ultimately the same as in the Duty Gear area; the only difference is that in Duty Gear, the material used is more complex and more rigid, such as thick plastic, while in Bianchi, the material used is softer and softer, such as leather or hide.

Automotive: Among the products offered by the company is the automotive sector, which is dedicated to making specific assemblies designed for automobiles, such as seat covers, steering wheel protectors, tire protectors, and assembly of accessories for the same approach. The main sections were thermoforming, sewing, and packaging.

4.2.1 Statement of the Problem

Within the range of existing products in the Body Armor area, vests are in great demand owing to their design, comfort, support, and protection. These vests are the most widely produced in the area and the most requested by customers.

The demand for vests is greater than what can be processed, considering the number of workers who can report to work during a pandemic. The problem arose due to orders accumulated during the quarantine period, which caused most people vulnerable to COVID-19 not to report to work. This period occurred during April and May 2020, when there was an increase in the backlog of orders to be produced.

Table 4.1 shows a comparison of production versus demand during the first half of 2020, as well as the efficiency of the finished product. It can be observed that during the period from January to June, only in the first two months (January and February) was the production demand met, while in April and May, the lowest production efficiency was obtained. The production standard is 24 pieces per shift.

The drop in production in April and May was caused by the unfortunate death of personnel, even counting the days of overtime offered to personnel who continued to work during that month. Table 4.2 shows the number of days worked per month on vest production lines in both shifts. It also shows the

TABLE 4.1 Comparison of Production and Demand in the First Half of 2020

MONTH	*PRODUCTION*	*DEMAND*	*EFFICIENCY*
January	2,150	2,000	107.50%
February	2,064	2,000	103.20%
March	1,978	2,000	98.90%
April	1,290	1,800	71.67%
May	1,204	1,500	80.27%
June	1,764	2,000	89.60%
Total	10,450	11,300	92.48%

number of active lines and the number of workers who showed up for work. It can be seen that April and May were when the least number of people showed up to work. Consequently, the number of overtime days increased.

Currently, 42 personnel process the vests. Previously, there were 56 people, representing a decrease of 14. Each production line comprised 12 sewists, an assembler, and an inspector. The plan is to set up a specialized line to process the models of vests in the most significant demand and adapt the line operators to work with similar models so that they become familiar with them and the learning curve is steeper. This new line takes advantage of the resources of the two existing lines to form a single line, and the available resources include machinery, people, and space. The space planned to be used is where lines 13 and 14 are located, as these lines are where the vest models are processed. Figure 4.1 shows the current layout of these lines.

TABLE 4.2 Days Worked During the First Half of the Year 2020

MONTH	*DAYS WORKED*	*DAYS OF OVERTIME*	*LINES ACTIVE*	*PERSONS WORKING*
January	22	3	4	56
February	20	4	4	56
March	20	3	4	56
April	22	8	2	28
May	20	8	2	28
June	22	6	4	42

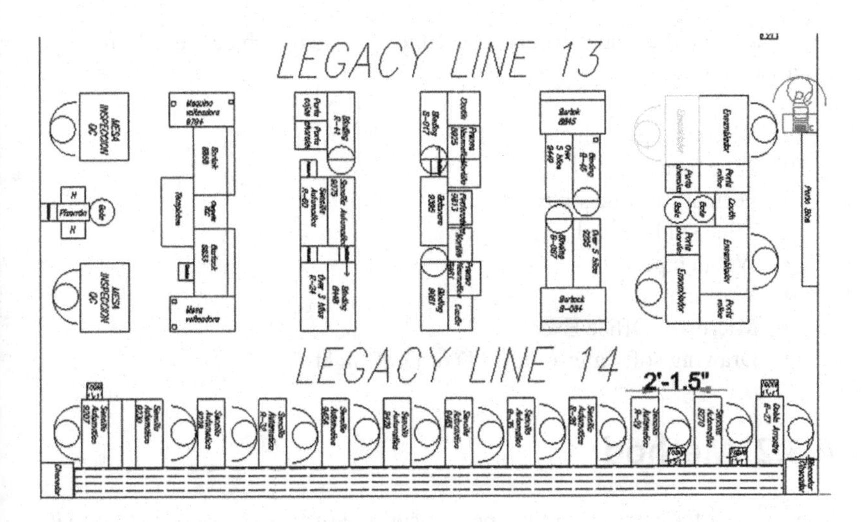

FIGURE 4.1 Current layout of production lines.

During June, line 13 had seven operators (six sewists and one assembler), while line 14 had 14 operators (12 seamstresses, one assembler, and one inspector); the first and second shifts were in the same state. The prospect is to set up a single line, taking advantage of the personnel resources of both shifts to achieve a higher production than the current one because, with the current number of personnel, it is impossible to do so.

4.2.2 Objectives

The general objective of this case study was to satisfy the production demand with the currently available human resources. On the other hand, the specific objectives are the following:

- Achieve the goal of 100% efficiency in the monthly production of vests.
- Reduce overtime days to 0.
- Relocate 14 people to the staff.

4.3 MATERIALS AND METHODS

4.3.1 Materials

The materials, tools, and programs used to achieve the objectives of this project are as follows:

- Digital stopwatch
- Tape measure
- Format for time study
- Time study board
- Wooden pencil
- Worktables
- Microsoft Office Excel
- Drawing software: AutoCAD or DraftSight

4.3.2 Method

The strategy for carrying out the project consists of five phases of the DMAIC methodology: defining, measuring, analyzing, implementing, and controlling.

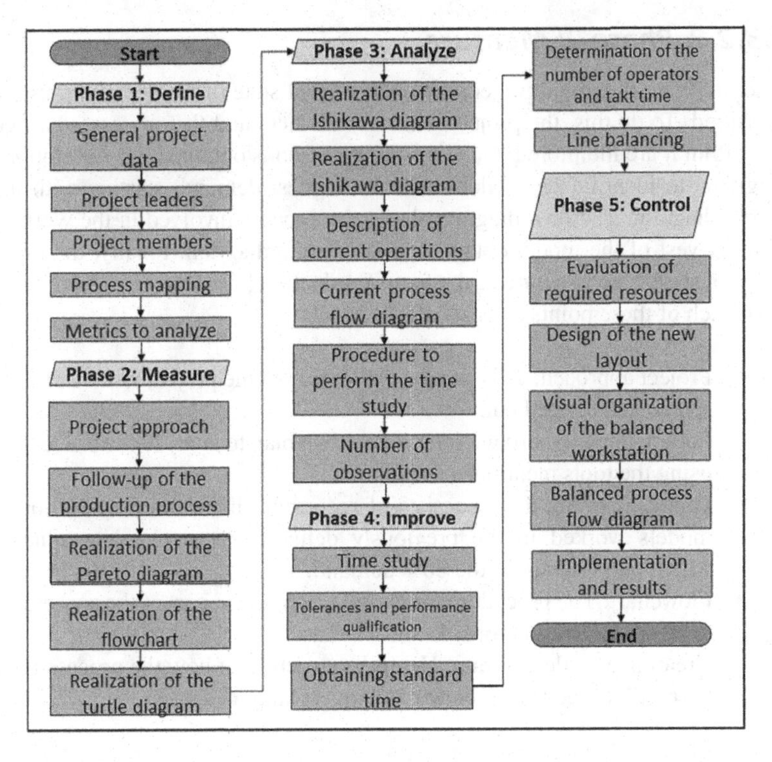

FIGURE 4.2 Methodology to be used in the project.

Figure 4.2 shows the breakdown of each phase as well as each of the activities to be performed. The following subsections show how each method was applied.

4.3.2.1 Phase 1: Define

In this phase, the general data of the project are established, the project leaders are defined, and all the members are involved. A process mapping is also carried out, where the input and output of the material are projected from the moment the client requests it, and the metrics to be analyzed are defined. The following is a description of each point.

- General Project Data: A summary of the project is provided.
- Project members: Each support team member involved was listed.
- Process mapping: To elaborate on the products made by the company, it first shows the needs customers want to satisfy, focusing specifically on the variety of items that the company itself can offer.
- Metrics to be analyzed: The process evaluation points are shown.

4.3.2.2 Phase 2: Measure

After defining the initial project data, the current state of the demand must be measured. To do this, the project is clearly defined, and the processes carried out within it are monitored. Once this information is obtained, we develop the diagrams to identify the models with the highest demand, starting with the Pareto diagram. The flow diagram shows the process involved in the work of a single vest of the model obtained in the Pareto diagram. Finally, the turtle diagram shows the structure of the focus production line. Next, it is mentioned what each of these points consists of.

- Project approach: This section explains how the project will be presented and carried out.
- Follow-up of the production process: We plan to analyze the process using the tools mentioned further on.
- Pareto diagram: This consists of recording the series of parts, or models worked in the previously defined process to determine which parts represent the 80% demand.
- Flowchart: The procedure for making a complete part of the process to be studied was recorded.
- Creating a turtle diagram: Here, the structure of how the process is currently in the line to be worked on is elaborated.

4.3.2.3 Phase 3: Analyze

In this phase, an Ishikawa diagram is constructed to determine the root cause of the problem. Next, the process is analyzed, and each operation is described for a better approach. Finally, a route diagram was constructed, showing the distances and paths covered by each operator. The method used to begin the study must be defined to review the current status. The following is a description of each point.

- Ishikawa diagram: The causes of the main problem are analyzed to emphasize opportunities for improvement.
- Analysis of the current process: The document currently handled by the operators and line leaders is analyzed to determine the activities to be carried out by each responsible. The heading indicates the vest's model and style, year of creation, sample number, production target, update date, and image of the vest design.
- Description of current operations: This section describes the breakdown of each operation carried out in the process.
- Current process route diagram: The layout shows the route taken by the operators during each operation within the line.

- Procedure for time study: The time study was carried out using the time-taking board, a wooden pencil, a digital stopwatch, and the time-taking form on which the model was to be analyzed, the date of the study, and the name of the analyst.
- Calculation of the number of observations: The sample size of the time taken was calculated using a statistical method. This method states that preliminary timings should be taken, and with these data, calculate n with a confidence level of 95.45% and a margin of error. n' preliminary time series, and with these data, calculate n with a confidence level of 95.45% and a margin of error of 5%. For this purpose, Equation 4.1 was used (Salazar-López 2024).

$$n = \left(\frac{40\sqrt{n'\sum x^2 - \left(\sum x\right)^2}}{\sum x} \right)^2 \tag{4.1}$$

Where:
n = sample *size* to be calculated (number of observations).
n' = number of observations in the preliminary study.
x = value of observations.

4.3.2.4 Phase 4: Improve

In this phase, most of the project's work is carried out, as this is where the changes and improvements being obtained begin to be observed. First, a time study is carried out, considering the observations calculated at the previous point and the established procedure. In addition, the tools mentioned in the Materials section must be available, including a digital stopwatch, a board for time study, and a sheet with the format for recording the time taken. Once the times were obtained, they were captured in Microsoft Office Excel software, and formulas were generated to consider the required tolerances and clearances, thereby obtaining the standard time. As a last point for this phase, a line balancing is performed according to the number of operators required and the takt time obtained in the time study. Each point can be explained as follows:

- Time study: This shows the operations and the duration of each one of them within the process.
- Tolerances and performance qualification: The clearances considered to evaluate the process and operators are defined.

- Obtaining the standard time: The average time study data and clearances were obtained.
- Determination of the number of operators and takt time: The total number of operators involved in the new process and the reasons for this are defined.
- Line balancing: The new breakdown of operations is performed with the number of operators estimated at the previous point.

4.3.2.5 Phase 5: Control

The resources needed for the new production line were evaluated in this last phase. Then, the current layout area was measured with a tape measure, and the new line was designed using DraftSight drawing software. Additionally, work tables or instructions can be added to each operation. Once the line is finished, an FMEA analysis is performed to detect any inconveniences. Once the line was balanced, the implementation was carried out, and during the project's application, the improvements that could be made were recorded until the expected results were obtained. The following describes how each of these points is carried out.

- Evaluation of required resources: The human and machinery resources required to create a new production line are shown.
- Design of the new Layout: A new production line is drawn using the estimated resources.
- FMEA Analysis: Possible failure modes along the line were recorded.
- Visual organization of the balanced workstation: The design of how the information will be displayed to each operator so that they know the operation to be performed is determined.
- Balanced Process Path Diagram: The new layout shows each operator's new path.
- Implementation and Results: The production results recorded during a specific period are shown.

4.4 RESULTS

4.4.1 Results of Phase 1: Define

4.4.1.1 General Project Data

The data described in. what follows provides brief and general project information.

- General objective: To meet production demands with the currently available human resources.
- Specific objectives: Achieve 100% efficiency in the monthly production of vests, reduce overtime days to zero, and relocate 14 people to the staff.
- Justification: Efficiency and production output are improved in models with the highest demand.

4.4.1.2 Project Team

The team in charge of carrying out the project consisted of ten members. Table 4.3 shows the team members by position and activity performed by each team.

4.4.2 Results of Phase 2: Measure

Table 4.4 shows the monthly production demand for each vest model with its corresponding percentage. It can be seen that model 1 was the one that obtained the highest demand, with 1,162 units, representing 46.97%, while model 2 obtained a demand of 824 units, equivalent to 33.31%. Note that these first two models accounted for 80.27% of the total demand, which is why the present study focuses on these two models. Figure 4.3 shows a Pareto diagram of the demand for the vest models.

On the other hand, Figure 4.4 shows the flow diagram obtained from the activities and the path that must be followed to process a vest of models 1 and 2. The orange ovals represent the start and exit of the vest manufacturing process; the yellow parallelograms represent the sections through which the vest passes and the blue rectangles represent the activities carried out.

TABLE 4.3 Project Team Members

POST	*ACTIVITY*
Project leader	In charge of the main activities.
Project sub-leader	Making decisions in each activity.
New products	Provide samples of the required models.
Area Manager	Supervise and approve internal processes.
Planning	Assignment of orders to process line.
Finance	In charge of continuous improvement projects and savings review.
Maintenance	Provide support to machinery and layout movements.
Security	Plant health and safety manager.
Supervisor	In charge of line production.
Line Leader	Line staff leader.

TABLE 4.4 Demand for Vest Models on September 11, 2020

MODEL	*QUANTITY*	*%*	*ACCUMULATED*
1	1,162	46.97%	46.97%
2	824	33.31%	80.27%
3	127	5.13%	85.41%
4	164	6.63%	92.04%
5	53	2.14%	94.18%
6	32	1.29%	95.47%
7	31	1.25%	96.73%
8	22	0.89%	97.62%
9	15	0.61%	98.22%
10	13	0.53%	98.75%
11	12	0.49%	99.23%
12	11	0.44%	99.68%
13	8	0.32%	100.00%
TOTAL	2,474	100%	

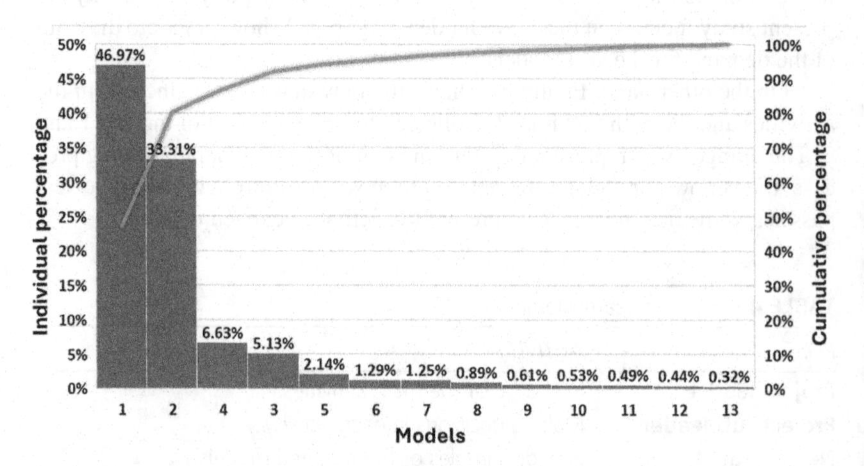

FIGURE 4.3 Pareto diagram of the demand for vest models.

4.4.3 Phase 3 Results: Analyze

Figure 4.5 shows an Ishikawa diagram for the problem of low production. Different causes were detected, including a lack of personnel due to the pandemic.

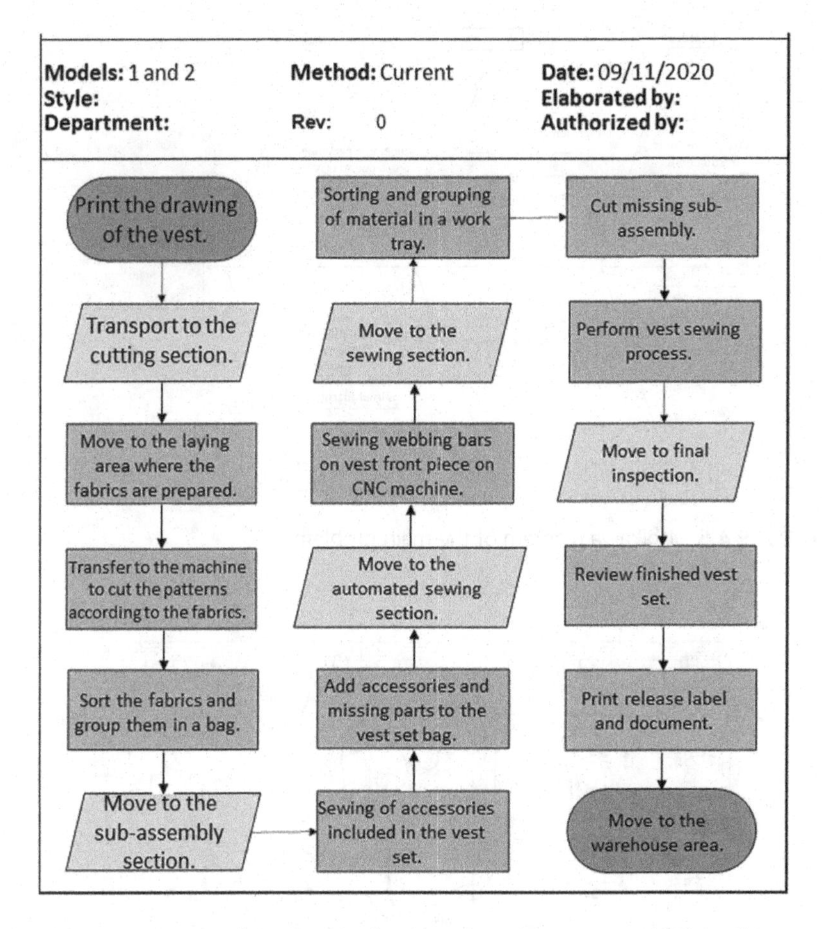

FIGURE 4.4 Process flow diagram for manufacturing vests models 1 and 2.

Figure 4.6 shows the initial routing diagram of the current process used in the production lines. The process begins when the allocator delivers the material to the assembler. Afterward, the material is prepared in a work tray, the sub-assembly is cut, added, and finally processed on the production line until it reaches inspection, where the process is validated, and the complete set of parts is assembled and sent to the warehouse.

Table 4.5 shows the time study with five preliminary samples.

By applying Equation (4.1), $n = 3.21$ was obtained. This is almost equal to 3. Because the process duration is relatively long and there are few variations between the preliminary times, the result indicates that only three repetitions should be carried out for the time study.

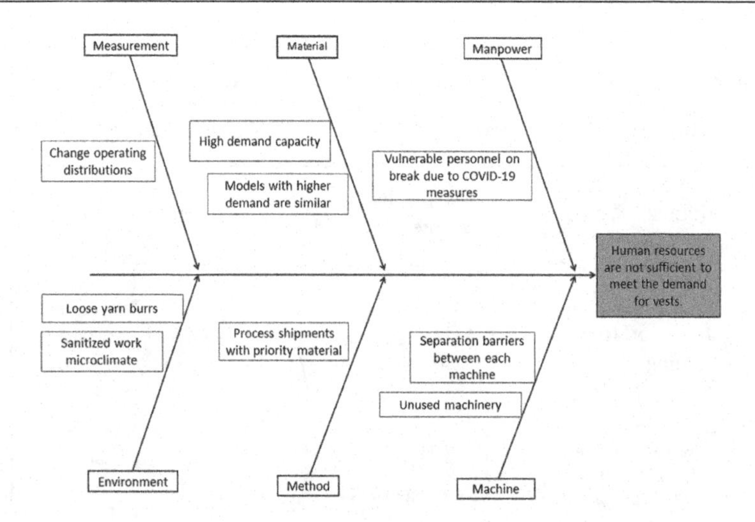

FIGURE 4.5 Ishikawa diagram of the main problem.

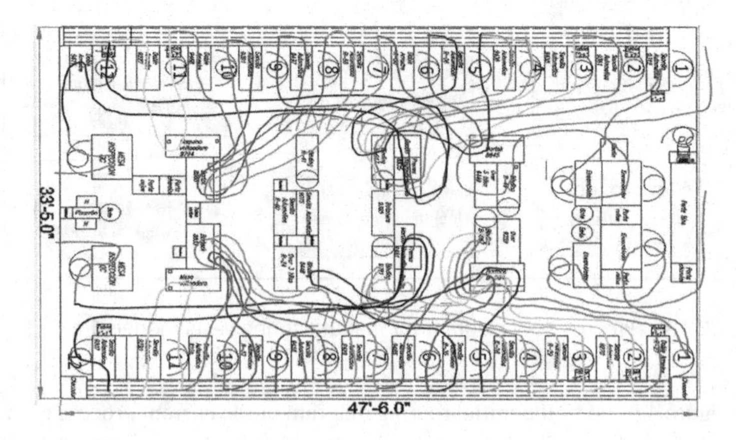

FIGURE 4.6 Initial routing diagram of production lines.

4.4.4 Results of Phase 4: Improve

To carry out the first study, emphasis was given to Model 2 because this model has more operations than Model 1. Therefore, the study was used to apply a balance to both models. The manufacturing process of Model 2 comprised 12 operations, which comprised between eight and 20 subtasks. Therefore, Table 4.6 shows only the average sum of the times of the sub-tasks of each

TABLE 4.5 Preliminary Timing Results

REMARKS	*X (MINUTES)*	X^2
1	192.15	36921.6225
2	176.55	31169.9025
3	184.37	33992.2969
4	169.68	28791.3024
5	188.65	35588.8225
Total	911.4	166463.9468

operation of the production process of Model 2. When evaluating the work pace of the operators during the process, it was considered that a 15% tolerance should be added to the time obtained because other factors affecting time were detected, such as repairs, lack of material, cleaning of the area, ergonomic exercises, and shift start-up meetings. The performance qualification factor was considered to obtain the standard time, and the operator's performance at the time of the activity was evaluated. The Westinghouse system was used to rate the performance. Once the tolerances and performance qualifications were considered, these values were added to the average time initially obtained, and the standard times per operation shown in Table 4.6 were obtained.

The takt time obtained was 1,007.29 seconds, corresponding to operation 7. With this takt time, a cycle time (runtime) of 3.92 hours is obtained. This time is required to process the vest of model 2. Likewise, with this takt time, a maximum production output of 31 pieces is obtained, which means seven more pieces than what is required by the standard.

As mentioned, there are currently 42 people on the staff, so we had the option of setting up three lines of 14 total personnel. However, setting up another line in either shift is impossible because the staff is split between the two shifts. Therefore, a more extensive line was selected to exploit these individuals. The line used half of the current human resources (21 people), divided between 18 sewers, two assemblers, and one inspector. The initial lines of 12 sewists have a production goal of 24 pieces per shift; therefore, the lines of 18 sewists would have to produce their proportional equivalent, that is, 36 pieces.

However, taking advantage of the results obtained in the time study and eliminating the subtasks of the assembly of waist elastics and the construction of the IDs of the front part, which made operations 8 and 12, respectively, it was estimated that the production would be 48 pieces (31 pieces with 12 seamstresses + 15.5 pieces with the six added sewists = 46.5 pieces + 1.5 pieces of the operations to be taken out of the process = 48 pieces); that is, the same as two lines of 12 sewists would produce, meaning 50% more daily production.

TABLE 4.6 Time Study for Model 2

OPERATIONS	*AVERAGE TIME (SECONDS)*	*STANDARD TIME (SECONDS)*
Assemble the back cover	776	928.1
Assemble the front cover of the combo	765.5	924.34
Make upper front buttonholes	811.5	939.81
Make front-side zipper bags	832.5	982.16
Assemble the entire back apron piece	510.5	622.3
Assemble the back base piece with velcro	656	792.12
Make front-side buttonholes	922	1,007.29
Attach back suspension	822	954.75
Fasten breastplates to the front	637.5	826.19
Assemble the front gun bag base piece	832	956.8
Close and flip the sword	849	971.78
Turn over the small front	648.5	932.08

Figure 4.7 shows the layout for the design of the new line.

With that study and the planned goal of 48 parts per 8.9-hour shift, it was estimated that the cycle time per part output would be every 11.13 minutes. Each vest had a standard process time of 5.1 hours. They are multiplying 5.1 hours by the expected 48 pieces; 244.8 hours of production gains are expected. The number of operators would increase to 21, with 18 sewists, two assemblers, and one inspector. Multiplying the 21 operators by 8.9 hours of work yielded a total of 186.9 hours gained. Dividing the 244.8 hours earned by the 186.9 hours paid gives a total of 130.98%, which is higher than the current efficiency of 98.23 %, which means an increase of 32.75% (see Table 4.7).

The production times with the 18 sewing machines are shown in Figure 4.8 and Figure 4.9 for models 1 and 2, respectively. In the case of model 1, it can be seen that the highest time was for sewist 7, with a time of 535.9 seconds, while the lowest time was for sewist 14, with a time of 503.7 seconds. This gave a difference of only 32.2 seconds, which represents a reduction of 91.64% between the highest and lowest times compared to the initial configuration of the production line.

For model 2, it can be seen that the longest time was for sewing machine 5, with a time of 608.35 seconds, while the shortest time was for sewing machine 17, with a time of 571.55 seconds; that is, there was a difference of 36.8 seconds between the two. This represents a decrease of 90.44% between the longest and shortest times compared to the initial configuration of the production line. In addition, with the new configuration, the maximum times were reduced by 46.8% and 39.61% for Models 1 and 2, respectively, compared to the initial configuration.

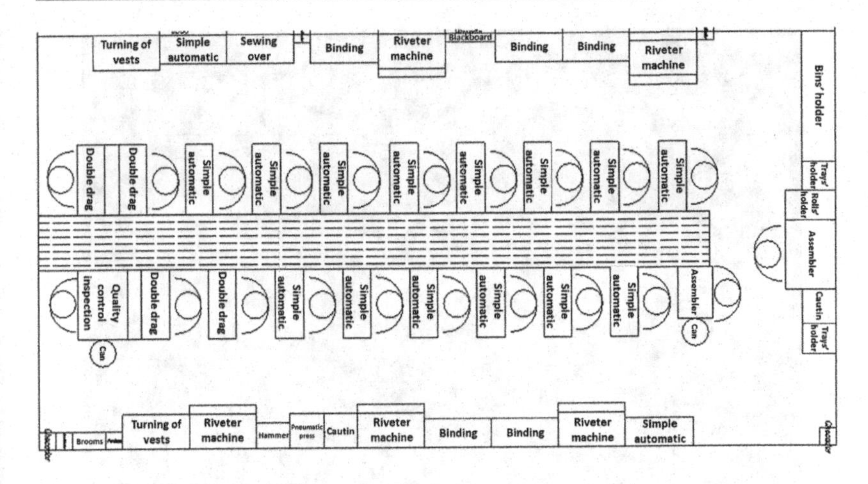

FIGURE 4.7 Proposed layout of the new production line.

TABLE 4.7 Data Obtained with the Information from the Time Study of Model 2

INDICATOR	*VALUE*
Cycle time (seconds)	667.8
Parts expected per day (target)	48
Hours paid per piece	5.1
Expected hours earned per production	244.8
Operators (18 seamstresses, two assemblers and one inspector)	21
Hours worked per operator	8.9
Paid hours per line	186.9
Efficiency	130.98%

4.4.5 Results of Phase 5: Control

The FMEA analysis showed that out of 59 subtasks, only one presented a significant risk, with a risk priority number (RPN) value of 315. The remaining subtasks had an RPN of less than 280.

Regarding the visual organization of the workstation, layout sheets of Models 1 and 2 were obtained for their respective documentation, as well as to inform the line leaders how the new process flow will be. Finally, so that each seamstress was familiar with their operation, work instructions were designed for each operation and placed in their respective workstations. Figure 4.10 shows a sample of the work instructions for Operation 1 of Model 2.

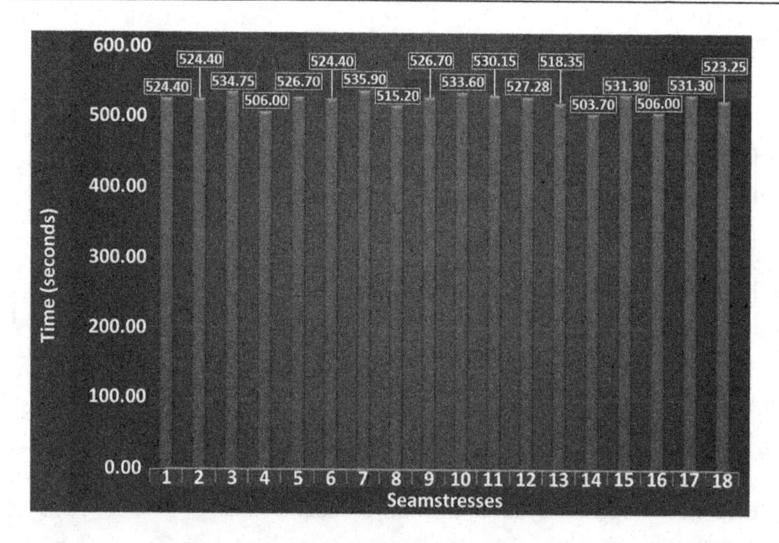

FIGURE 4.8 Line balancing of model 1.

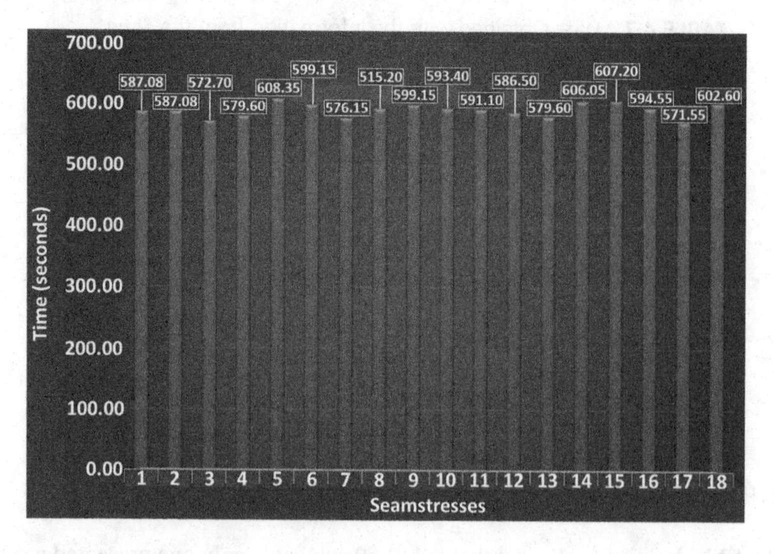

FIGURE 4.9 Line balancing of model 2.

Figure 4.11 shows the initial and final path diagrams with the new layout design for the balancing process. It can be seen that there was a significant difference between the routes taken by each of the operators before and after the new layout design. This new design begins with the delivery of material to the assembler, where the fabric separation and sub-assembly preparation processes are distributed. It then goes to the production cell for the sewing process

MODEL 2

OPERATION 1

MOLLE CUMMERBUND (1 SIDE):

- **To make a bastilla for molle cummerbund lining.**
- **Attach velcro hook to molle lining.**
- **Make bastilla to molle cummerbund.**
- **Mark and attach 4" velcro loop. Make X seam.**
- **Attach lining to molle cummerbund.**
- **Pique molle cummerbund and turn to the front.**
- **Overstitch to molle cummerbund.**
- **Mark and pin edges of front web bars.**
- **Attach molle cummerbund and molle web edges to front.**
- **Rivet molle cummerbund bar joint to front and front bar edges.**
- **Clean chalk on the front.**

FIGURE 4.10 Sample work instructions for Operation 1 of Model 2.

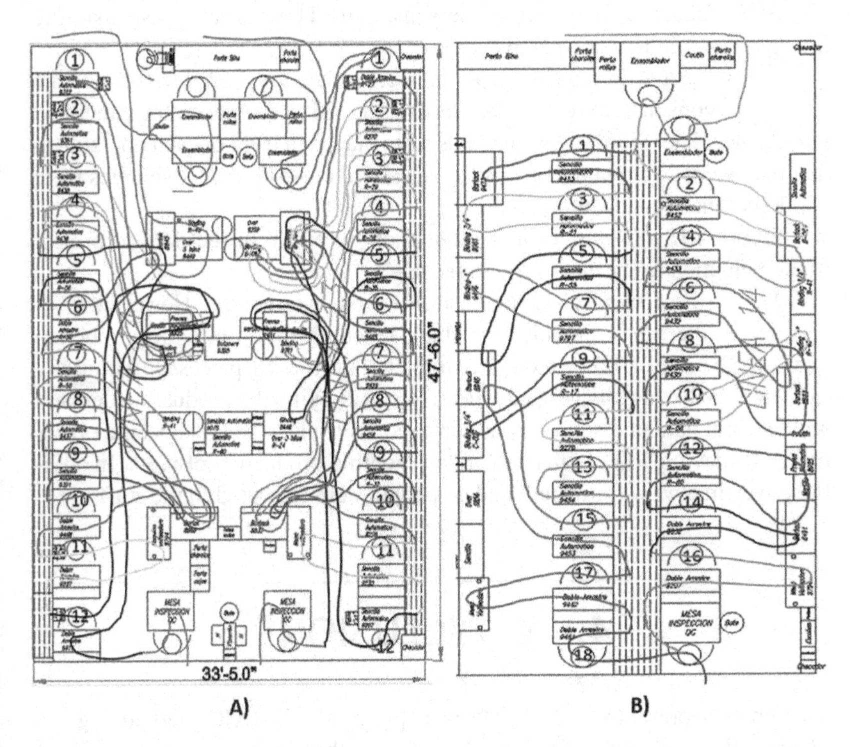

FIGURE 4.11 Production line routing diagram: A) initial, B) final.

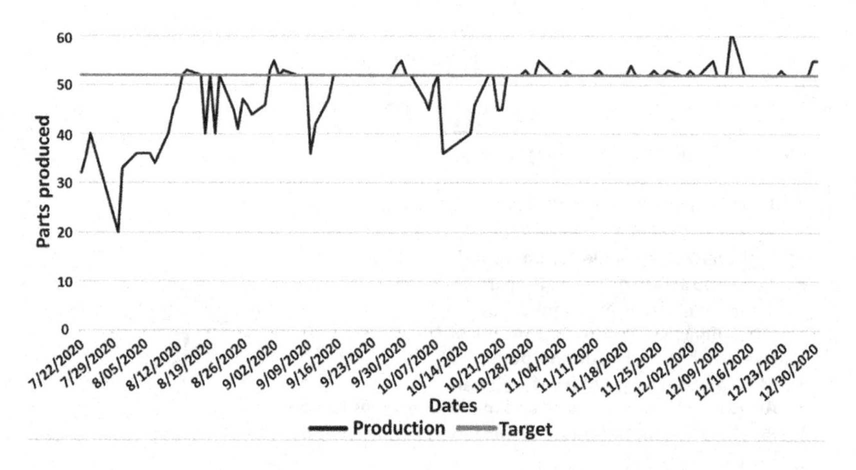

FIGURE 4.12 Production of parts made during the second half of 2020 on the new production line.

with the support of several machines. Afterward, it goes to the quality check station for validation, and finally, a person assigned by the warehouse takes the finished material to its respective area.

The project was implemented on Wednesday, July 22, 2020. The line leader was consulted to review the places assigned to each operator integrated into the new line. Similarly, visual aids were placed in front of each operator so that they were aware of the process to be carried out. The same dynamics were applied to the second shift.

Figure 4.12 shows the line's production output from the date of project implementation to the end of the same year.

There was a trend of variation in the first two months. However, as of November, the trend remained constant, reaching the minimum production goal of 52 pieces/shift, which was higher than the 48 pieces/shift initially expected. With this, the general objective of satisfying the production demand with current human resources was achieved, as well as the first specific objective of achieving 100% efficiency in the monthly production of vests. In addition, overtime was eliminated, and 14 people were relocated to the staff.

4.5 CONCLUSIONS

This chapter presents a case study of applying the DMAIC methodology to increase the production of two models of vests through redesigning a production line. Concerning the literature, the results of this chapter are similar to those obtained by Mohamad et al. (2018). These authors analyzed the problem of defective parts in a company that manufactures electronics, where they applied the DMAIC methodology. Consequently, the company significantly reduced the product rejection rate and improved customer satisfaction. In another study, a mechanical component manufacturing plant faced a problem of low production. To solve this problem, Hamza (2008) implemented the DMAIC methodology to improve the cycle times of production processes. In his study, this author identified bottlenecks and downtime using tools such as the Ishikawa diagram and process capability analysis and implemented actions to eliminate them. Consequently, a notable reduction in the production cycle time and an increase in the plant's overall production capacity were obtained.

These case studies, together with those presented in this chapter, illustrate how the DMAIC methodology can be successfully applied in different industrial contexts to achieve significant improvements in efficiency and productivity. In all cases, the implementation of the DMAIC methodology has a positive impact on companies. The thorough identification of problems in production processes and implementation of appropriate corrective actions result in significant improvements in efficiency and productivity.

Furthermore, it is essential to note that applying the DMAIC approach not only solves immediate problems, such as rejecting defective products or optimizing production cycle time but also leads to cultural change within organizations. Commitment to continuous improvement and the ability to proactively identify and address problems has become an integral part of corporate culture. These case studies also highlight the importance of the DMAIC methodology as an effective tool for data-driven decision-making. Accurate data collection and analysis enable teams to identify critical areas for improvement and objectively evaluate the impact of implemented actions.

REFERENCES

Armijal, Marlina WA, Hadiguna RA (2023) The evaluation of supply chain risk management on smallholder layer farms. IOP Conf Ser Earth Environ Sci 1182:012082. https://doi.org/10.1088/1755-1315/1182/1/012082

Boyle TA, Scherrer-Rathje M (2009) An empirical examination of the best practices to ensure manufacturing flexibility: Lean alignment. J Manuf Technol Manag 20:348–366. https://doi.org/10.1108/17410380910936792

Cai M, Luo J (2020) Influence of COVID-19 on manufacturing industry and corresponding countermeasures from supply chain perspective. J Shanghai Jiaotong Univ 254(25):409–416. https://doi.org/10.1007/S12204-020-2206-Z

Clegg CW, Wall TD, Pepper K, et al. (2002) An international survey of the use and effectiveness of modern manufacturing practices. Hum Factors Ergon Manuf Serv Ind 12:171–191. https://doi.org/10.1002/HFM.10006

Hamza SEA (2008) Design process improvement through the DMAIC Six Sigma approach: A case study from the Middle East. Int J Six Sigma Compet Advant 4:35–47. https://doi.org/10.1504/IJSSCA.2008.018419

Harvey J, Lefebvre LA, Lefebvre E (1992) Exploring the relationship between productivity problems and technology adoption in small manufacturing firms. IEEE Trans Eng Manag 39:352–358. https://doi.org/10.1109/17.165417

Herron C, Braiden PM (2006) A methodology for developing sustainable quantifiable productivity improvement in manufacturing companies. Int J Prod Econ 104:143–153. https://doi.org/10.1016/J.IJPE.2005.10.004

Kapoor K, Bigdeli AZ, Dwivedi YK, et al. (2021) How is COVID-19 altering the manufacturing landscape? A literature review of imminent challenges and management interventions. Ann Oper Res 2021:1–33. https://doi.org/10.1007/S10479-021-04397-2.

Kumar A, Luthra S, Mangla SK, et al. (2020) COVID-19 impact on sustainable production and operations management. Sustain Oper Comput 1:1–7. https://doi.org/10.1016/J.SUSOC.2020.06.001

Mohamad N, Ahmad S, Samat HA, et al. (2018) The application of DMAIC to improve production: Case study for single-sided flexible printed circuit board. In: Ratnam MM (ed.) IOP Conference Series: Materials Science and Engineering. IOP Publishing, pp. 1–11.

Okorie O, Subramoniam R, Charnley F, et al. (2020) Manufacturing in the time of COVID-19: An assessment of barriers and enablers. IEEE Eng Manag Rev 48:167–175. https://doi.org/10.1109/EMR.2020.3012112

Salazar-López B (2024) Industrial engineering Online.com: Calculation of the number of observations: Sample size in the time study. In: Ingeninería Ind. Online.Com: Cálculo del número Obs. Tamaño d ela muestra en el Estud. tiempos. https://ingenieriaindustrialonline.com/estudio-de-tiempos/calculo-del-numero-de-observaciones/. Accessed 21 February 2024.

Taqi HMM, Ahmed HN, Paul S, et al. (2020) Strategies to manage the impacts of the COVID-19 pandemic in the supply chain: Implications for improving economic and social sustainability. Sustainability 12:9483. https://doi.org/10.3390/SU12229483

Index

Printed in the United States
by Baker & Taylor Publisher Services

Printed in the United States
by Baker & Taylor Publisher Services